Julia Tomonko

Flood protection and the influence of forecasting systems

AF783655

Julia Tomonko

Flood protection and the influence of forecasting systems

An analysis of early warning models in Lower Austria and Upper Austria

ScienciaScripts

Imprint
Any brand names and product names mentioned in this book are subject to trademark, brand or patent protection and are trademarks or registered trademarks of their respective holders. The use of brand names, product names, common names, trade names, product descriptions etc. even without a particular marking in this work is in no way to be construed to mean that such names may be regarded as unrestricted in respect of trademark and brand protection legislation and could thus be used by anyone.

Cover image: www.ingimage.com

This book is a translation from the original published under ISBN 978-3-330-51421-8.

Publisher:
Sciencia Scripts
is a trademark of
Dodo Books Indian Ocean Ltd. and OmniScriptum S.R.L publishing group

120 High Road, East Finchley, London, N2 9ED, United Kingdom
Str. Armeneasca 28/1, office 1, Chisinau MD-2012, Republic of Moldova, Europe
Managing Directors: Ieva Konstantinova, Victoria Ursu
info@omniscriptum.com

Printed at: see last page
ISBN: 978-620-2-37480-4

Copyright © Julia Tomonko
Copyright © 2025 Dodo Books Indian Ocean Ltd. and OmniScriptum S.R.L publishing group

Table of contents

CHAPTER
Background and questions

In this thesis, the author deals with the topic of "Flood protection and the influence of forecasting systems". She aims to explain the importance of systematic and comprehensive flood protection, which is usually only recognised by all parties after major floods have occurred and in view of the consequential damage. In Austria, the recent major flooding on the Danube has focused attention on flood risk management. The authorities responsible for flood risk management include the respective municipalities and district authorities. They are increasingly being called upon to change their existing flood protection policies, to request other flood protection methods and, ultimately, to implement them. The focus has shifted from purely technical flood protection structures to non-technical management strategies. Since technical measures such as the construction of dams have been the main focus in the past, but these often reach their limits due to ever-increasing flow rates, the aim now is to achieve modern, integrated flood risk management. This aims to combine various preventive and management measures. In addition to spatial planning and the creation of retention areas, great importance is attached to flood forecasting and early warning systems. Two types can be distinguished here. On the one hand, short-term warnings are issued, which are used specifically for operational purposes due to their high forecast accuracy. On the other hand, longer-term early warnings are intended to increase the preparedness of the population and emergency services immediately before flood- s (cf. BLÖSCHL 2008: 41).

Since it is often difficult to provide sufficient space for the construction of technical flood protection structures such as dams in densely populated areas, but effective protection against flooding from rivers such as the Danube is still desired, flood forecasts are becoming increasingly important.

Technical flood protection is also often unfavourable for tourism and landscape design reasons. For this reason, mobile flood protection systems are already in use in many municipalities and cities (cf. BLÖSCHL et al. 2013: 9).

Increasing importance is also being attached to property protection, personal responsibility on the part of affected citizens, willingness to recognise risks and thus also to reduce damage through the active participation of the population. In order to keep the extent of damage as low as possible, timely evacuation or property protection is required, which in turn requires a functioning early flood warning system. A central aspect of modern, integrated flood management is the effective interaction of all these measures (cf. BLÖSCHL et al. 2013: 10).

Research and projects make an important contribution to incorporating new findings into the complex hydrological models used for flood forecasting. The aim is to improve the accuracy of flood forecasts and to enable more reliable flood protection planning. Mobile flood protection systems, for example, require a long preparation time (cf. BLÖSCHL et al. 2013: 10).

Although the timing of an impending flood cannot be predicted weeks in advance due to the complexity of the processes involved, flood risk management allows for precise preparation for flood events (cf. BLÖSCHL et al. 2013: 11).

Furthermore, powerful computer technologies have greatly expanded the availability of satellite data, fast data transmission networks and global weather forecasts, thereby greatly advancing forecasting technology. However, these methods have also increased political and public pressure on hydrographic services. As more and more private and commercial properties are located near rivers and therefore in floodplains, the value of the area increases, as does the extent of damage in the event of flooding, and there is a greater interest in obtaining reliable and timely forecasts (cf. BLÖSCHL 2008: 41).

In recent years, there have been repeated floods on the Danube, with extremely high flow rates recorded in some sections, causing devastating damage in parts of Upper and Lower Austria. The general assumption on which this work is based is that absolute protection against flooding is neither feasible nor sensible. Therefore, timely and reliable warning of approaching flood hazards is becoming increasingly important. The research question can therefore be formulated as follows: What are the functions of flood forecasting systems and early warning systems? The hypothesis is that computer forecasts and weather models can be used to increase the warning time and thus reduce damage (cf. BMLFUW 2002: 33).

It is assumed that flood events will continue to occur due to the complex interaction of various influencing factors ( ). Since the Danube has been exposed to numerous anthropogenic environmental factors in the past and climatic changes are increasingly leading to more intense precipitation, the question of human responsibility repeatedly arises when searching for the causes of extreme events. The next research question is therefore: How much does climate change contribute to the increased occurrence of floods? The corresponding hypothesis is that climate change leads to an increase in extreme weather conditions, which have already triggered numerous flood events in Austria in the form of precipitation (cf. BLÖSCHL 2015: 13).
Many Austrian municipalities, especially those in the Wachau region, extend close to the riverbed due to intensive settlement. From a tourism and spatial planning perspective, a mobile flood protection system is the only option for these municipalities. The final research question is therefore: What factors

are necessary for a functioning mobile flood protection system? The corresponding hypothesis is: Mobile flood protection only works in conjunction with an early flood warning system (cf. OFFICE OF THE LOWER AUSTRIAN PROVINCIAL GOVERNMENT 1996).

The definitions and explanations of the basic terms necessary for this thesis and intended to prepare for the subsequent empirical part are dealt with in chapters four to eight of this thesis. This section also examines the research questions "How much does climate change contribute to the increased occurrence of flooding" and "What factors are necessary for mobile flood protection to work?" Furthermore, the corresponding hypotheses are verified with sufficient literature.

The empirical part consists of qualitative guided interviews and an interpretation of these interviews with regard to the research questions and hypotheses posed in order to be able to test the theoretical assumptions in practice. This section is specifically dedicated to the research question "What functions do flood forecasting systems and early warning systems have?" The hypothesis posed is also verified here.

Since the citation guidelines of the University of Vienna (Institute of Geography and Regional Research) have been applied in full, no references are provided in the thesis. For example, italics indicate intellectual property protection.

CHAPTER 2

Summary

In Austria, floods are the most frequent and damaging natural disasters. In recent years, flooding of the Danube in particular has caused the most economic damage in Austria (cf. BLÖSCHL 2015: 3).

Floods are natural events that do not occur evenly, but are divided into periods with little flooding and periods with high flooding. Unfortunately, floods cannot be prevented as they are part of the natural water cycle. Precipitation that cannot be retained by plants, soil or other natural means is discharged to the surface. Floods therefore occur when particularly unfavourable conditions that promote surface runoff coincide (cf. BLÖSCHL 2015: 3).

However, floods are also the result of human intervention, which influences runoff in many ways. Noteworthy examples include anthropogenic influences on the climate, the use of water bodies, areas close to water bodies and the redesign of watercourses to optimise their use. Essential parameters of the water balance, such as precipitation intensity, are therefore changing as a result of human activity (cf. BLÖSCHL 2015: 3).

Due to the use of areas close to water bodies, attempts must be made to use the available design options to

Achieving optimum protection. In many local conditions, including in the Wachau region, it is not possible or desirable to construct permanent flood protection measures ( ). Mobile protection systems that work in close cooperation with flood forecasting systems are therefore used (see BMLFUW 2002: 33).

A flood forecast must therefore perform the highly complex task of combining various factors such as weather, vegetation, soil and landscape forms in order to be able to make a realistic forecast on the one hand and to control flood protection activities by humans on the other (cf. Blöschl et al. 2013: 9).

CHAPTER 3

Theoretical foundations

This chapter presents basic definitions, classifications and technical knowledge on the subject of floods.

3.1 Runoff vs. flow

Figure 1 shows the difference between discharge and flow, which is that in flow, the volume flow per unit of time only flows through a surface water cross-section. The volume flow per unit of time that leaves the catchment area above and below ground through the discharge cross-section is referred to as discharge (cf. DYCK and PESCHKE 1983: 33).

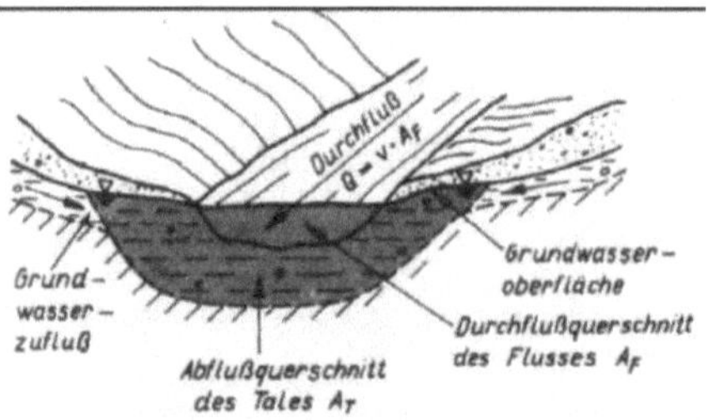

Fig. 1: Explanation of the terms discharge and flow (source: DYCK and PESCHKE 1983: 33, Figure 4.6)

3.2 Water level measurement

The water level indicates the height of the water surface above a fixed reference horizon at a specific location and is measured using gauges. Different gauging systems are used for measurements in the Danube region. Figure 2 shows a manually operated slat gauge. This is the simplest form and is used as a supplement to all mechanically operated gauging systems (cf. DYCK and PESCHKE 1983: 49).

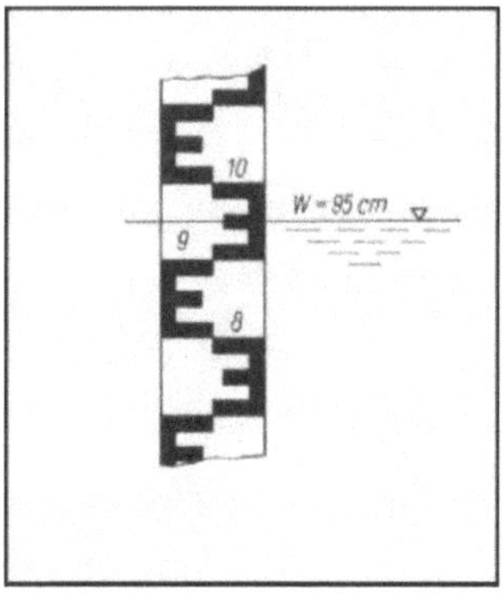

Fig. 2: Water level gauge (source: DYCK and PESCHKE 1983: 49, Figure 7.2)

4.2.1 Equipment of the water level measuring stations on the Danube

As shown in Figure 3, the water level measuring stations on the Danube are equipped with both water level gauges and remote data transmission.

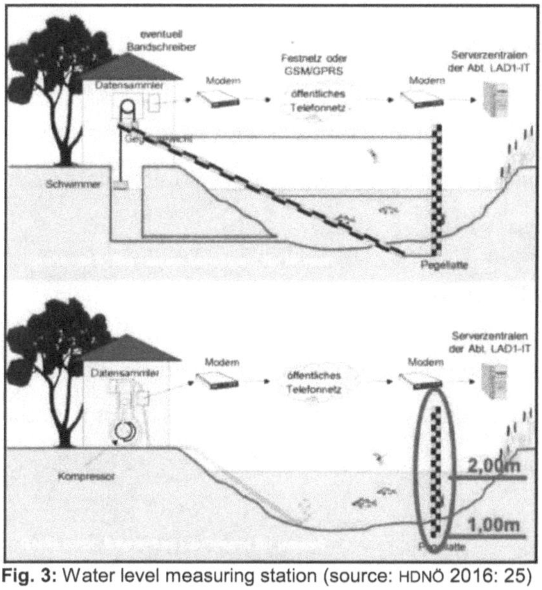

Fig. 3: Water level measuring station (source: HDNÖ 2016: 25)

The remote data transmission system automatically transmits the measured values to the hydrological service of the respective state government. Monitoring personnel are required to check that the values calculated automatically by the data collector installed at the measuring station correspond to those displayed manually on the gauge. The tasks of this monitoring personnel therefore include reading the water level from the gauge and passing on the values to those responsible at the hydrological services. The measured values are compared in a specialised model operated by the respective hydrological services of the federal states and described in more detail in Chapter 7.2 (cf. HD NIEDERÖSTERREICH 2016: 25).

4.3 Flood

According to THE EU FLOODS DIRECTIVE (2007: 27ff.), flooding is a temporary inundation of land that is not normally covered by water. Flooding is a natural phenomenon that cannot be prevented and includes, for example, flooding caused by rivers.

DYCK AND PESCHKE (1983: 328) define flooding as follows:

> "A flood (HW) is a temporary increase in flow above the base flow [sic!] that exceeds a limit (e.g. flood discharge) determined for each flow cross-section [sic!] determined from statistics or local conditions (e.g. flood discharge [sic!]), as a consequence of meteorological events or events caused by disasters."

3.3.1 Classification of floods

Flood events can be classified based on their "annuality". The term "annuality" describes the probability of recurrence of a flood event with the associated discharge volume (PATT 2001: 573).

Description of HQ($_{Tn}$) (= Tn-annual flood) according to ÖNORM B 2400 (2004: 5):

> "In an infinitely long series of observation years, the Tn-annual flood is reached or exceeded on average every $_{Tn}$ years. This information

does not allow *the time when this event will occur to be determined."* e.g. HQ100, HQ30, etc.

The "annuality" is determined in the Danube study area on the basis of centuries of observation and statistical evaluation of historical flood events (see BMVIT 2016).

Table 1 illustrates the return period statistics for the Danube.

Design case	Recurrence	Discharge (m^3 /second)
RHHW	Calculated Maximum high water	14,000
HQ 100	100-year flood	11,200
HQ 30	30-year flood	10,05
HQ 10	10-year flood	7,400
HQ 5	5-year flood	6,650
HQ2	2-year flood	5,800
HQ 1	1-year flood	5,150

Table 1: Annual recurrence statistics for the Danube (data source: VALENTI 2006: 23; own representation)

The highest discharges since 1402 (values from the Kienstock/Wachau gauge) are shown in Table 2.

Design case	Year/date	Discharge (m3/second)
HQ 100	4 June	11,450
HQ	24 June 2009	8,200
HQ 100	14 August 2002	11,305
HQ 10	23 March 2002	8,588
HQ 10	4 August 1991	9,647
HQ 10	1985	7,300
HQ 10	1981	7,440
HQ 10	2 July 1975	8,800
HQ 10	1965	7,520
HQ 30	13 July 1954	10,200
HQ 10	1920	7,980
HQ 100	17 September 1899	11,200
HQ 10	2 August 1897	9,900
HQ 10	9 June 1892	8,340
HQ 10	1890	7,770
HQ 10	4 January 1883	8,520
HQ 30	4 February 1800	10,500
HQ 100	1787	11,900
RHHW	1501	14,000
HQ 30	1402	10,500

Table 2: Historical flood events (data source: BMVIT 2016; own representation)

3.3.2 Flood parameters

The following parameters can be subjected to statistical analysis, as listed in Table 3.

Parameters	Definition
Flood level [cm, m]	Water level above a fixed reference horizon.
Flood peak discharge [m^3/s]	Maximum discharge during the passage of a flood wave in a body of water.
Rise velocity [cm/h]	Speed of the water rise.
Flood duration [h]	Begins at the time of wetting with water and ends at the time of drying.
Flood hydrograph	Graphical representation of values in chronological order.
Flow velocity [m/s]	Decisive intensity parameter.
Runoff coefficient	Quotient of the discharge height of the direct runoff (effective precipitation) and the precipitation height of the associated precipitation event.

Table 3: Flood parameters (data source: BMLFUW 2011: 17, own representation)

3.3.3 Types of flooding

PATT (2001: 6f.) distinguishes between the following types of floods:

- flash floods,
- storm surges and
- river floods.

Flash floods occur suddenly, either after a local heavy rainfall event or after a dam burst or the breach of another barrier (landslide), causing large masses of water to rush into an area (cf. PATT 2001: 7).

Storm surges occur on sea coasts. They are caused by hurricane-like winds that drive the water towards the coasts. This leads to a considerable rise in the water level on the coasts (cf. PATT 2001: 7).

River floods do not normally occur unexpectedly. Floods in river basins occur during prolonged, widespread rainfall or snowmelt. The size of the catchment areas and their characteristics influence the changes in water levels (cf. PATT 2001: 7).

CHAPTER 4

Introduction to the study area

Two large-scale geological developments were decisive for the formation of the Danube. The formation of the Alps was triggered by the northward drift of the African continental plate and is therefore directly linked to the retreat of the primordial sea towards the east. Previously, however, many different water systems with a west-east gradient shaped the present-day Danube landscape (cf. JUNGWIRTH et al. 2014: 50f.).

The present-day Danube is believed to have originated around 10 million years ago in the Krems area of Austria, where several watercourses first converged. From here, the "Ur-Danube" carved its path and began its slow but steady advance westwards. Over time, the watercourses that formed its catchment area also increased in volume. The reversal of the flow direction was possibly triggered by a meteorite impact. As a result of uplift, the Danube river system, which had developed from many water systems, could no longer flow westward and was thus forced to find a new course. About 6 million years ago, after the sea receded, the Danube finally flowed into the Vienna Basin (cf. KEILER and THALLER 2009: 44).

4.1 Geographical overview

With a total length of 2,850 km, the Danube is the second longest river in Europe after the Volga in Russia. It rises in neighbouring Germany east of Donaueschingen (Baden-Württemberg) and flows into the Black Sea via a multi-armed delta. The Danube's total catchment area ( ) covers an area of approximately 817,000 ^{km2}. Austria accounts for about 10% of this area. Figure 4 illustrates the river system of the Danube and its three main sections. The upper Danube stretches from its source in the Black Forest to the breakthrough through the mountain range consisting of the foothills of the Alps, the Leitha Mountains and the Little Carpathians. The middle Danube then flows to the Iron Gate, the gorge in the southern Carpathians. The section from the Iron Gate to the mouth, which encompasses the entire delta region, is known as the lower Danube (cf. JUNGWIRTH et al. 2014: 26ff.).

On Austrian territory, the Danube begins at Passau and ends after 357 kilometres between the Hundsheim Mountains and the Little Carpathians. It therefore flows through two federal states, Upper Austria and Lower Austria. This section of the river is classified as the Upper Danube and has a total gradient of 155 metres. With this gradient, the "Austrian" Danube has the character of a mountain river and reaches flow velocities of 1 m/s to 3 m/s (during floods) (cf. KEILER and THALLER 2009: 43; JUNGWIRTH et al. 2014: 16).

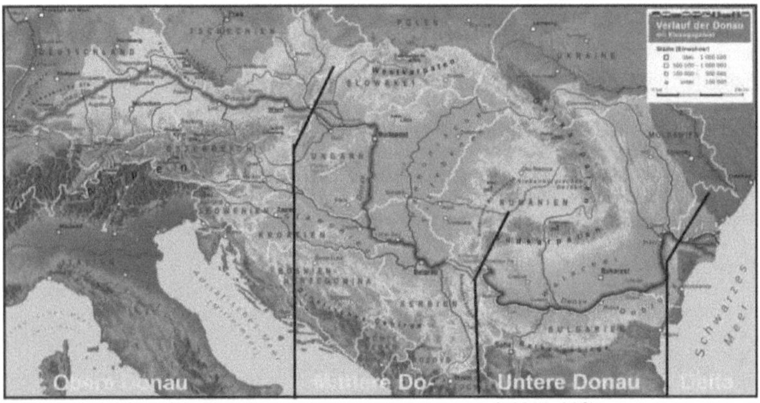

Fig. 4: Subdivision of the river system (source: JUNGWIRTH et al. 2014: 18f.)

4.2 Geology and morphology

According to GRILL (1950: 10), the landscape character of the Danube Valley, as shown in Figure 5, is determined by the geological structure of the respective river area. Between Passau and Krems, it is, on the one hand, the southern foothills of the Bohemian Massif, the geological structure of the Mühlviertel and Waldviertel regions, where the river carves its way through the rocks, and, on the other hand, the geological widening of individual bays of the Tertiary Alpine foreland basin with its clayey, marly and sandy sedimentary rocks, which influence the geological structure (cf. GRILL 1950: 10).

Between Krems and Vienna, in the so-called Tullner Basin, the Danube reaches the edge of the Alps and separates the youngest and outermost area, the flysch or sandstone zone (cf. GRILL 1950: 10).

The small, young Tertiary Korneuburg Basin lies in the middle of the Vienna Gate and causes the river landscape to widen (cf. GRILL 1950: 10).

From the Vienna Basin, the Danube enters the large subsidence area within the Alpine-Carpathian arc, which was formed in the Neogene period.

Along the river, one encounters many geological differences in the overall structure of Austria. The Bohemian Massif, the Alps and the Tertiary Basin are the defining geological units of the country of Austria (cf. GRILL 1950: 10).
Morphologically speaking, the Austrian Danube region can be characterised as a sequence of narrow valleys and basins in which the Danube ( ) forms wide floodplains with typical floodplain vegetation (cf. JUNGWIRTH et al. 2014: 120 ff.).

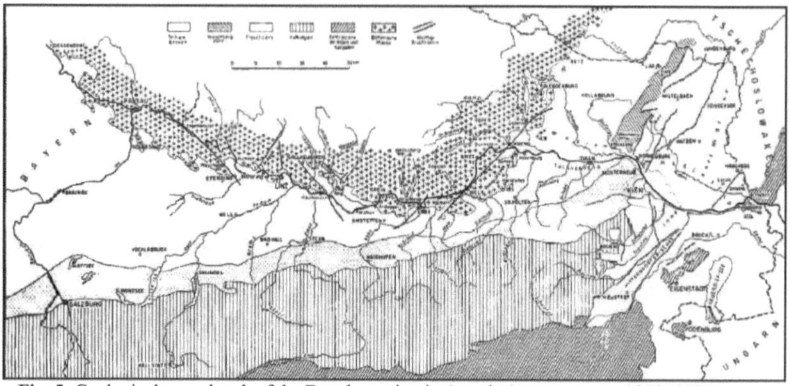

Fig. 5: Geological map sketch of the Danube region in Austria (source: GRILL 1950: 459)

4.3 Climate

The Austrian Danube basin is located in temperate latitudes and has a Central European transitional climate characterised by oceanic and continental influences. As shown in Figure 6, Austria is divided into four further climate provinces based on its geographical location, surface shape and altitude. The Danube lies in the transitional climate zone of Upper Austria. This area is located north of the Alps and is influenced by the Atlantic Ocean. As a result, there are neither extremely hot summers nor extremely cold winters, and it rains more frequently here than in other parts of Austria. In Lower Austria, however, the Danube is located in the Pannonian climate zone. Here, the climate is hardly influenced by the sea. It rains less and the temperature differences between summer and winter are greater than in the rest of Austria (cf. FREYTAG-BERNDT and ARTARIA KG 2008: 25).

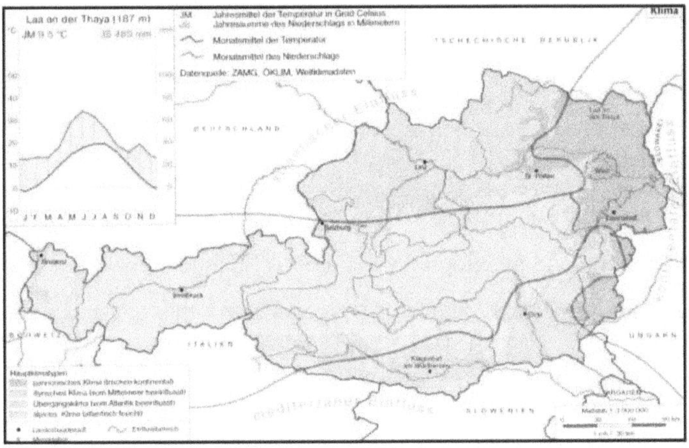

Fig. 6: Main climate types (source: FREYTAG-BERNDT and ARTARIA KG 2008: 25)

CHAPTER 5

Factors influencing flooding

In principle, the formation of floods can be divided into the two main groups "hydroclimatic" and "physiographic". The most important factors resulting from the hydroclimatic and physiographic characteristics for the course of a flood are initial conditions, area characteristics and event characteristics (cf. NIEHOFF 2001: 3).

Various hydrological reservoirs such as retention space, soil storage or basin storage can be mentioned for a rough characterisation of the initial conditions. Important functions of these reservoirs with regard to flood formation are their general capacity, their capacity at the beginning of the event and their filling and emptying behaviour. The loss of retention space along the river and in its catchment area leads to increased runoff and thus to accelerated flood peaks (cf. NIEHOFF 2001: 3).

Area characteristics refer to the physiographic factors that influence runoff formation. These characteristics include anthropogenic interventions such as soil compaction due to intensive land use or land sealing due to urbanisation. Sealing the soil surface is the greatest intervention in natural processes in terms of runoff formation. Soil compaction is caused, among other things, by heavy agricultural machinery. This process is the result of intensive use and reduces the storage capacity of the soil (cf. NIEHOFF 2001: 5).

Hydroclimatic factors are characterised by various events such as precipitation or snowmelt. The intensity and duration of a precipitation event have a strong influence on how much the characteristics of the area shape the course of flooding (cf. NIEHOFF 2001: 3).

The following subchapter examines the hydroclimatic aspect in detail.

5.1 Influence of weather on flood formation

The significance of certain weather situations and the occurrence of extreme weather events in Central Europe was recognised very early on, around 1891, by Wilhelm J. Van Bebber. He conducted extensive research in this area, evaluated low-pressure areas over Europe and defined typical trajectories of low-pressure areas (cf. HOFSTÄTTER et al. 2015: 21).

Climatological observations have shown that weather conditions associated with heavy precipitation will continue to occur. When these precipitation events coincide with other unfavourable conditions that promote surface runoff, devastating floods can occur in Central Europe. Changes in precipitation or in the risk of flooding are being accelerated by anthropogenic and natural climate change, among other factors. Higher temperatures and the associated higher absolute humidity of the air, as well as changes in general circulation and changes in low-pressure activity, are examples of such changes. This can also influence the frequency and duration of certain circulation types or tracks and thus have an impact on precipitation patterns and the potential risk of flooding . The so-called "Vb-type" weather situation or track (Chapter 4.3) is an example of a prominent atmospheric circulation pattern that can be linked to some of the most severe flood events in Austria in recent decades (cf. HOFSTÄTTER et al. 2015: 21).

5.2 Determination of tracks

Low-pressure areas (cyclones) are spatially limited areas located in the lower atmosphere (ground level to approx. 12 km altitude). They can be distinguished from their surroundings by low air pressure. Compared to the surrounding air, the air mass in these areas has specific properties in terms of humidity, density and temperature conditions. Low-pressure systems are also dynamically active and associated with characteristic weather phenomena such as widespread precipitation. The air mass properties of cyclones are influenced in two ways. On the one hand, they are determined by the area over which the system develops and, on the other hand, they are also influenced by the constant exchange with the surrounding environment.

. The path that a cyclone follows during its life cycle is called its track (cf. HOFSTÄTTER et al. 2015: 85).

Specialised methods can be used to analyse such tracks, determining the centre and the spatial/temporal displacement of the cyclones. In addition to recording the displacement of a low-pressure centre, this study also takes into account so-called peripheral lows. Such lows are located on the periphery of large low-pressure complexes and generally move more quickly than the central low-pressure area (cf. HOFSTÄTTER et al. 2015: 85).

The track thus makes it possible to determine the position of the systems at any given time and to record the immediate area of influence in terms of weather impact in the study area. Furthermore, the track can be used to identify the geographical region from which a low-pressure area is moving towards Central Europe, thereby enabling the systems to be differentiated in terms of their characteristics (cf. HOFSTÄTTER et al. 2015: 85)..

5.3 Trajectory catalogue

HOFSTÄTTER (2015: 85) distinguishes between the following tracks:
ATL (Atlantic)
CON (continental)
MED (Mediterranean)
POL (polar or northern type)

TRZ (Central Europe)

Vb (Van Bebber's train Vb)

X-N (developing south of the Alps, initially migrating northwards) X-S (developing south of the Alps, initially migrating southwards)

5.3.1 Classification of migration paths

As shown in Figure 7a, depending on their origin, tracks that move into the TRZ (Track Recognition Zone 0.5°E-23.9°E, 42.3°N-56.2°N) are divided into different types. The Track Recognition Zone (TRZ) also includes the study area of this thesis. The tracks are divided into five main groups: ATL (Atlantic), STR (subtropical), MED (Mediterranean), CON (continental), POL (polar), as well as track type Vb ( ) and three further classes X-N, X-S and TRZ (cf. HOFSTÄTTER et al. 2015: 89).

The difference between the main groups and the further subcategories is that low-pressure areas of the types ATL, POL, CON, MED, STR (and partly Vb) move from outside into Central Europe and thus also into Austria, whereas the types TRZ, X-N and X-S develop within the zone marked in red (cf. HOFSTÄTTER et al. 2015: 89f.).

X-N or X-S type tracks must first be determined in the "nas*" or "med*" areas. These two areas are shown in Figure 7a, are located within Central Europe, south of the Alps, and have been given arbitrary names. Depending on the direction of movement, these systems are differentiated into the northward-moving type X-N or the southward-moving type X-S. Since low-pressure areas with high moisture content and a tendency to intense precipitation occur in the areas south of the Alps, these two types are treated separately. The TRZ track type is used for all low-pressure areas that develop within this area. This also includes all locally developed thunderstorm zones (cf. HOFSTÄTTER et al. 2015: 89f.).

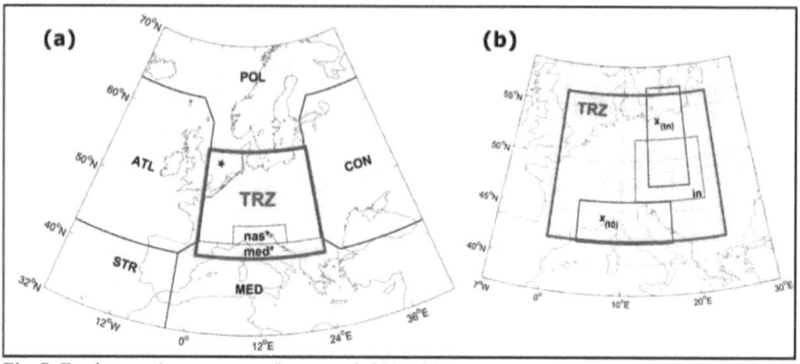

Fig. 7: Track types (source: HOFSTÄTTER et al. 2015: 90)

Figure 7b shows the track type "Vb". This is defined as a low-pressure area that moves from northern Italy – east of the Alps – northwards towards Poland. As this particular track type triggered the Danube floods of recent years with extreme precipitation events, it is analysed as a separate track class. A Vb track must meet the following criteria (cf. HOFSTÄTTER et al. 2015: 90):

- The system must be located in the area designated as "x(t0)" (4-17°E, 42-46°N) for at least a certain period of time.
- The system must then be located in "x(tn)".
- The system must have crossed the area "in" (12-22°E, 46-52°N) (cf. HOFSTÄTTER et al. 2015: 90f.).

The different migration paths and their areas of origin are shown in detail in Figure 8.

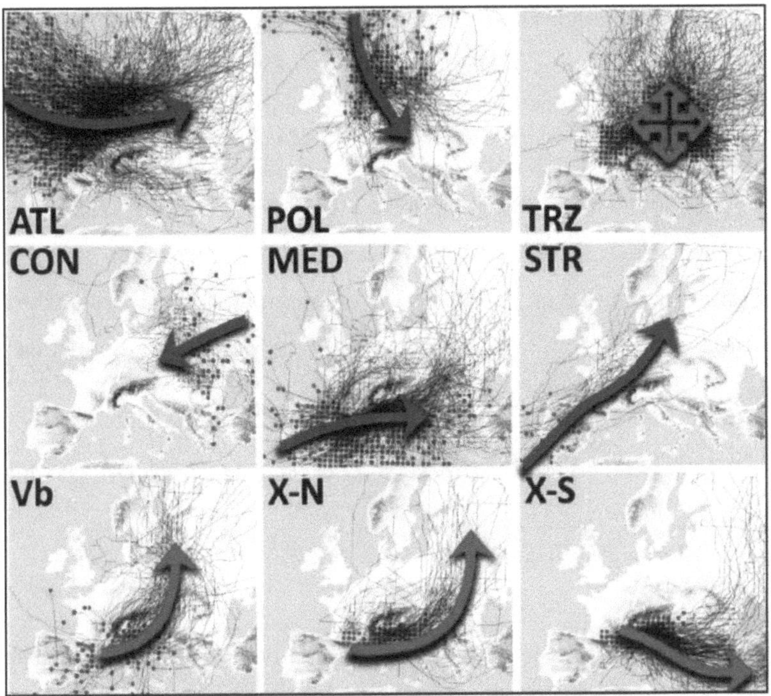

Fig. 8: Detailed representation of migration path types for Central Europe. The course of the individual migration path events is represented by the black line. The position of the origin of the migration path is indicated by a red dot. The arrow represents the typical course of the migration path. (Source: HOFSTÄTTER et al. 2015: 93)

The individual maps in Figure 8 show that each track type has characteristic patterns of spatial displacement of the tracks across Central Europe. It can also be seen that low-pressure areas pass through certain regions more frequently than others (cf. HOFSTÄTTER et al. 2015: 93).

> "These patterns are related, on the one hand, to the control of the systems in their displacement over the wind fields of the large-scale circulation and, on the other hand, to the underlying orography and surface" (HOFSTÄTTER et al. 2015: 94).

The dominant eastward shift of low-pressure areas from west to east is the most striking feature that can be observed and is naturally caused by the mean westerly drift of the general circulation. So-called "hot spots" of track frequency are the regions over northern Italy, the Ligurian Sea and the coastline between France and Denmark (cf. HOFSTÄTTER et al. 2015: 94).

5.4 Heavy precipitation - relevant tracks

Low-pressure areas in the Vb track are by far the most relevant for heavy precipitation. They are of utmost importance for the formation of major floods, although they occur extremely rarely. In conjunction with a so-called "cut-off low" in higher air layers, Vb tracks are more frequently found over Central Europe (HOFSTÄTTER et al. 2015: 8).

> *"A cut-off low is a low-pressure area that has a circular structure and has detached itself from the polar front"*
> (HOFSTÄTTER et al. 2015: 8).

This upper-level low is decisive for the long duration of precipitation events and high precipitation intensities. Vb tracks are particularly relevant in the Danube catchment areas during the summer months (May to October) (cf. HOFSTÄTTER et al. 2015: 8).

Another track with high precipitation relevance is the ATL type (cf. HOFSTÄTTER et al. 2015: 8).

5.4.1 Precipitation trends in Austria

Precipitation patterns in Austria are very diverse. The Alps play a major role in the distribution of precipitation. Precipitation is increased on the windward side of the mountains, while it is reduced on the leeward side ( ). In addition to the influence of the Alps, a west-east gradient in precipitation frequency can also be observed. The further east, the stronger the continental influence and the lower the precipitation. The various influences on heavy precipitation were investigated in a study (SEIBERT et al. 2006), which identified seven precipitation regions in Austria, shown in Figure 9. These localised areas can be assigned to weather conditions in which heavy precipitation occurs in the respective regions (cf. FORMAYER and KROMP-KOLB 2006: 2).

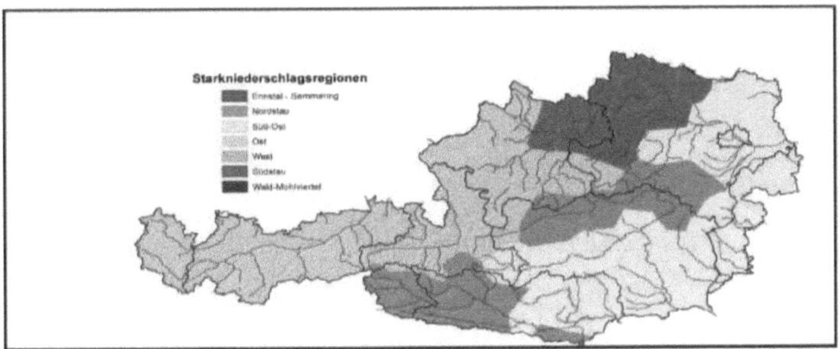

Fig. 9: Regions in Austria where heavy precipitation occurs simultaneously. (Source: SEIBERT et al. 2006)

5.5 Changes in the future climate

In general, a dipole structure can be observed in precipitation patterns in Europe. A decrease in precipitation is evident in the Mediterranean region and in Western Europe, including the British Isles and France. This decrease in precipitation is particularly pronounced in summer. In contrast, an increase in precipitation is expected in Scandinavia, the Baltic Sea region and Russia, which are classified as North-Eastern Europe. The Alpine region is in a transition zone. Although there is no clear precipitation trend here, a strong shift in precipitation across the seasons can be observed (cf. FORMAYER and KROMP-KOLB 2006: 3).

An increase in precipitation of 15-40% in the winter months is expected by the end of the century. The peak will be in February. In contrast, a decrease in precipitation of 10-50% is expected in the summer months. The peak will be in August (cf. FORMAYER and KROMP-KOLB 2006: 3).

With regard to Vb locations, the statements in climate change scenarios are not clear. In principle, the scenarios show a decrease in low pressure development in the Mediterranean region, which also leads to a sharp decrease in precipitation in the region. The frequency of Vb conditions should therefore also decrease. However, due to the warming of the Mediterranean Sea and the resulting evaporation, an intensification of precipitation in Vb weather conditions is very likely. In summary, with regard to Vb and Vb-like conditions, it can be stated that a decrease in frequency, currently around 10-15 events per year, but an increase in precipitation intensity per event is to be expected under these weather conditions. This would also increase the risk of flooding caused by these weather conditions. Figure 10 shows regions that are particularly affected by Vb and Vb-like conditions (cf. FORMAYER and KROMP-KOLB 2006: 6).

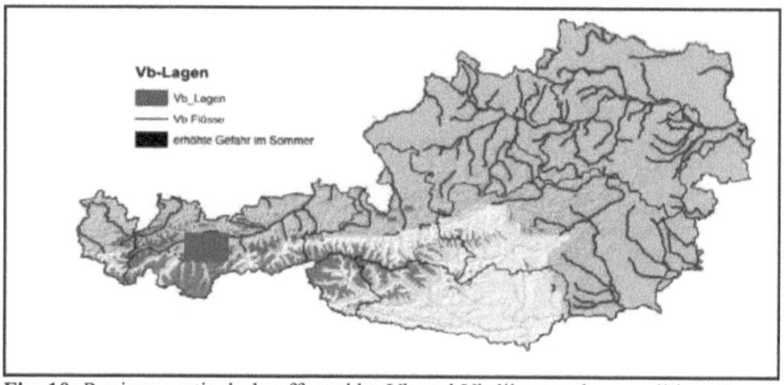

Fig. 10: Regions particularly affected by Vb and Vb-like weather conditions (source: FORMAYER and KROMP-KOLB 2006: 8)

> *"Flood events [...] can be triggered not only by precipitation fields of large-scale low-pressure areas, but also by short but intense heavy rainfall from thunderstorms"* (FORMAYER and KROMP-KOLB 2006: 8).

The Alps play a significant role in the frequency of thunderstorms. The Alpine foothills are where thunderstorms are most likely to occur. As can be seen in Figure 11, the Danube is not directly affected. However, it is exposed to several approaching thunderstorms (cf. FOR- MAYER and KROMP-KOLB 2006: 8).

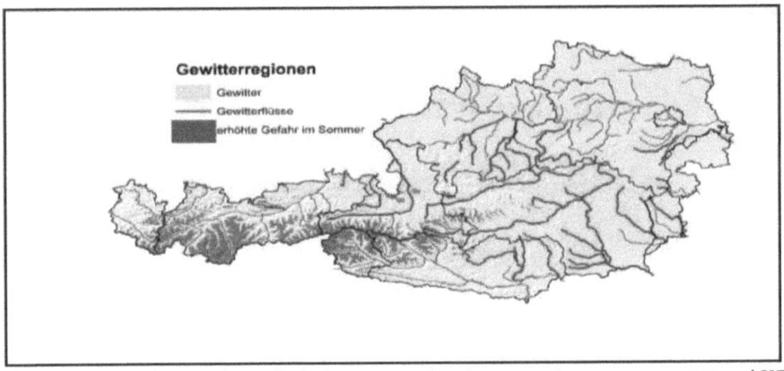

Fig. 11: Regions in Austria particularly prone to thunderstorms (source: FORMAYER and KROMP-KOLB 2006: 9)

CHAPTER 6

Forecasting and early flood warning

Flood forecasting is an essential component of preventive flood protection in the integrated flood management cycle (event, response, recovery and prevention). In order to coordinate the deployment of emergency services and technical flood protection measures effectively

, timely and spatially accurate forecasting is required (cf. GODINA 2006: 133).

> "Every analysis of major flood events highlights the importance of flood warning as an essential tool for risk and behavioural prevention and damage minimisation" (GODINA 2006: 133).

Over time, the requirements for flood forecasting have changed. A hundred years ago, an approximate forecast of the extent of the impending rise in water levels at a gauge was sufficient. However, today's integrated flood protection requires spatially and temporally accurate and early forecasting of discharge (cf. GODINA 2006: 133).

6.1 Development of flood warning systems in Austria

The water level and information service in the Austrian Danube catchment area, with the Imperial Hydrographic Department in Vienna as its central office, was established in June 1899. The water level information service has therefore been a central task of the hydrographic services of the provincial governments in Austria for more than 100 years ( ). The tasks of this service include the presentation and dissemination of current and historical water levels. The information service is based on data collected by the hydrographic measurement network. In addition to numerous other necessary input data, the data from the water level gauges are essential for the preparation of forecasts

and the early flood warnings derived from them. The development and operation of flood warning systems are therefore traditional tasks that have always been performed by the hydrographic services of the provincial governments in Austria (cf. GODINA 2006: 133).

Due to the numerous flood events in the last ten years, the importance of early warning has increased significantly, which has also accelerated the development of forecasting systems (cf. GODINA 2006: 133).

> "The requirements for mobile flood protection facilities on the Danube to provide accurate forecasts prompted the Lower Austrian Hydrographic Service *to switch* from an *empirical to a hydrological-hydrodynamic model* for water level forecasting and discharge forecasting for the Danube"
> (GODINA 2006: 133).

As can be seen in Figures 12 and 13, before 2002, flood forecasting systems existed only on the Danube, Salzach and Enns rivers. Due to requirements, more and more forecasting models for Austrian rivers have been developed over time or are still in planning ( ). These are shown in dotted lines in Figure 13 (GODINA 2006: 134).

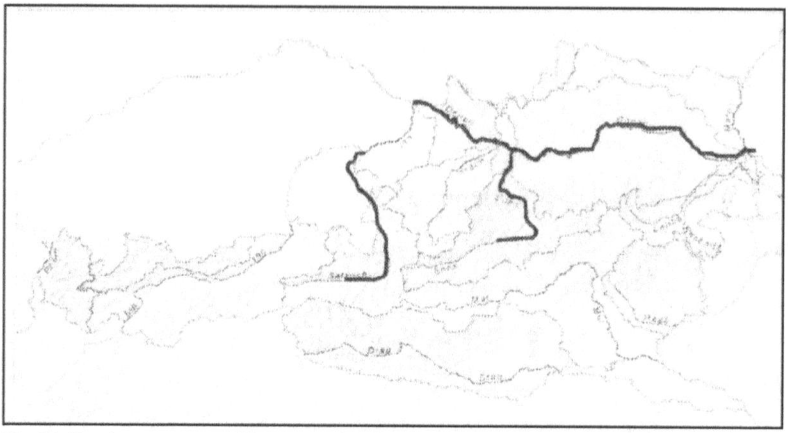

Fig. 12: Development status of forecast models in Austria in 2002 (source: GODINA 2006: 134)

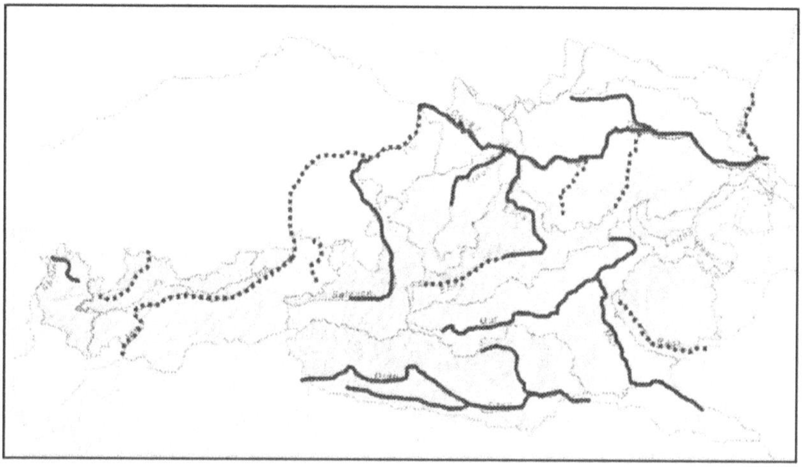

Fig. 13: Development status of forecasting models in Austria in 2006 (source: GODINA 2006: 134)

In addition to the hydrological model, a data management system is required, as shown in Figure 14, to manage the input values, as well as automated visualisation and distribution of information in the event of a predicted hazard, operated by the respective state warning centre (GODINA 2006: 135).

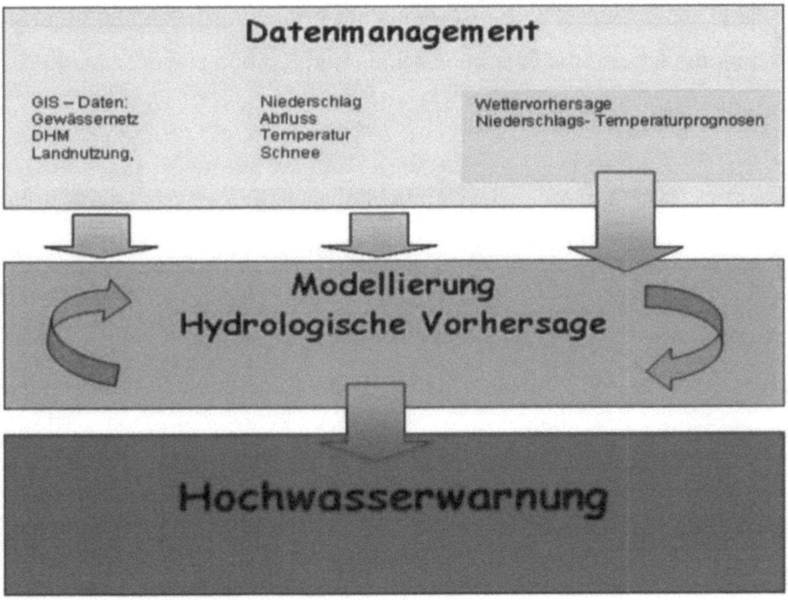

Fig. 14: Components of flood warning (source: GODINA 2006: 135)

In addition, precipitation forecasts from the Central Institute for Meteorology and Geodynamics (ZAMG), which are an important input for the hydrological models, must also be purchased. Hydrological models have developed enormously in recent years thanks to modern technology. In addition to observable measurement point information, secondary information

from the geoinformation system (GIS) is also used as input for modelling (GODINA 2006: 135).

The models are developed in cooperation with hydrographic services, university institutes, research centres and engineering firms. In order to meet the requirements for a functioning early flood warning system, cooperation between many experts from different disciplines is necessary. These disciplines include meteorology, hydrology, flood control, the State Warning Centre and IT. The cooperation covers data management in the respective internal IT environments, the development and use of the hydrological prediction model, and the comprehensible interpretation and presentation of the results (GODINA 2006: 135).

Another important part of the development of such forecasting systems is the analysis of historical flood events and the associated adaptation of the model parameters to the selected forecast model. In order to improve the forecasting of future events, the analysis and evaluation of the forecast results is also considered an essential factor in hydrological forecasting (GODINA 2006: 136).

The discharge on the Austrian Danube is also influenced by the operation of hydroelectric power plants. (cf. GODINA 2006: 135f.)

> "The extent of this influence on the course of a flood and its consideration in the hydrodynamic part of the models is important for the quality of the forecast results and requires that the control of run-of-river power plants and the release of reservoirs be taken into account in the forecast. The data and information required for this must be taken into account in the analysis of historical events and be available as input for forecasting" (GODINA 2006: 136).

6.2 Description of the flood forecasting system on the Danube

The flood forecasting system for the Danube calculates the expected discharge and water level for several forecast locations from measured hydrological data and forecast meteorological and hydrographic data. In this system, precipitation, temperature and discharges from several gauging stations are used as input data. In order to integrate the operation of the power plants on the Danube and its tributaries, specific power plant data is also fed into the system. Figure 15 provides an overview of how the expected discharges and water levels on the Danube are calculated from this input data (cf. SCIETEC 2006: 3).

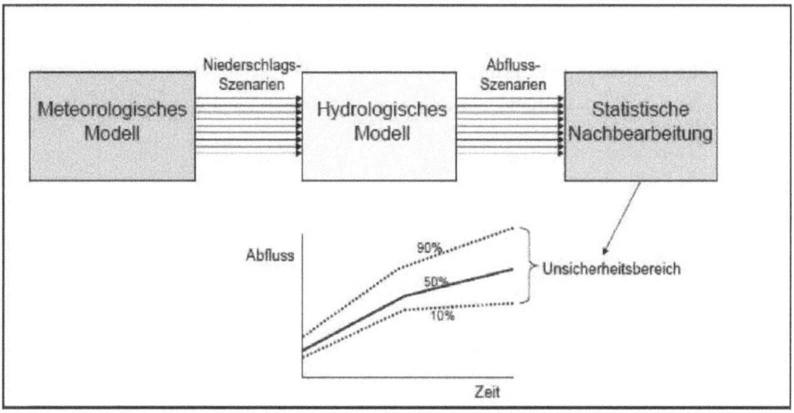

Fig. 15: Use of uncertainty information in flood early warning (source: BMVIT 2004: 23)

Another factor that must not be ignored is the current soil condition of the catchment area. This is defined by so-called state variables, such as soil moisture or snow cover. The current condition of the catchment area is determined either on the basis of previous calculations or by comparing the calculated and measured flows. Therefore, there is no direct measurement of soil moisture or similar (cf. SCIETEC 2006: 3).

The forecasting system on the Austrian Danube was developed as a continuously calculating system (cf. SCIETEC 2006: 3).

> "In a fixed cycle (1 hour), a calculation is automatically performed in which, on the one hand, the flow expected in the next 48 hours is calculated as Q(t) for the specified forecast points and, on the other hand, the new state in the catchment area (update from the last forecast time to the current time) is calculated" (SCIETEC 2006: 3).

Furthermore, as shown in Figure 16, the forecast system combines four different focal points (cf. SCIETEC 2006: 4):

- Precipitation forecasts (Fig. 16, green)
- Hydrological data (Fig. 16, dark blue)
- Hydrodynamic and hydraulic data (Fig. 16, light blue)
- Process control and visualisation (Fig. 16, yellow)

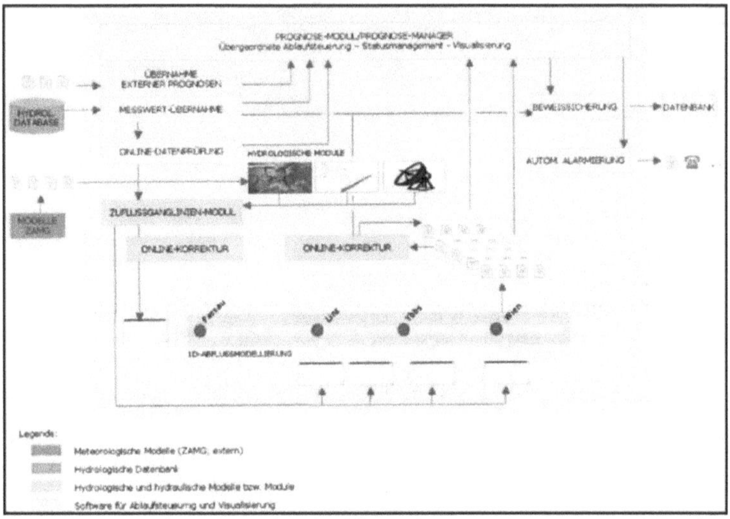

Fig. 16: Structure and data flows of the flood forecasting system (source: SCIETEC 2006: 4)

6.2.1Focus 1 – Precipitation forecasting

The first step in the flood forecasting model involves evaluating meteorological measurements obtained from stations operated by ZAMG, the Austrian Meteorological Service ( ), the Central Hydrographic Office (HZB) and Austro-Control. The locations of these measuring stations are shown in Figure 17 (cf. BMVIT 2004: 34).

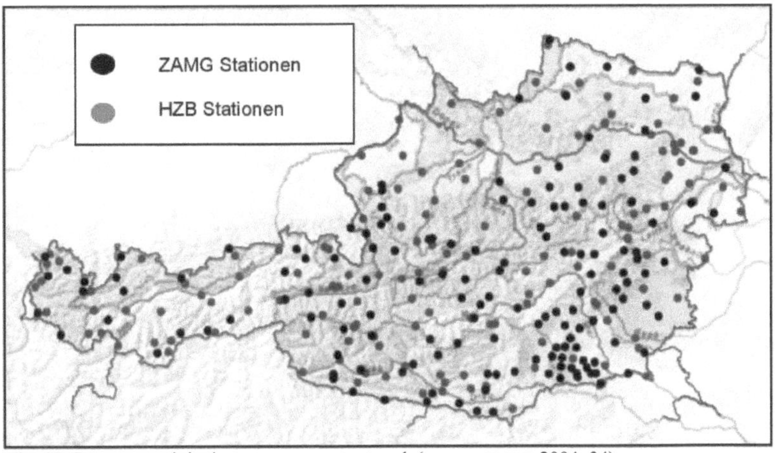

Fig. 17: Remote precipitation measurement network (source: BMVIT 2004: 34)

Precipitation and temperature forecasts for the next 48 hours are calculated using a meteorological model developed by ZAMG Vienna. The INCA system (INCA = Integrated Nowcasting and Comprehensive Analysis) specially designed for this purpose in Figures 18 and 19 contains meteorological measurement data, radar data and the calculations of two global circulation models from international meteorological institutions (ECMWF - European Centre for Medium-Range Weather Forecasts and ALA-RO/ALADIN - Aire Limitee Adaptation dynamique Developpement InterNational). With a resolution of 1x1 km, the system provides an analysis of precipitation and temperature forecasts (cf. OFFICE OF THE UPPER AUSTRIAN PROVINCIAL GOVERNMENT 2013: 2).

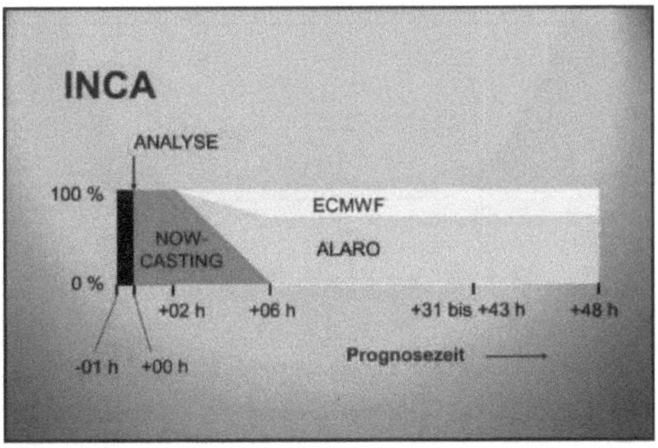

Fig. 18: INCA system (source: HD NÖ 2016: 32)

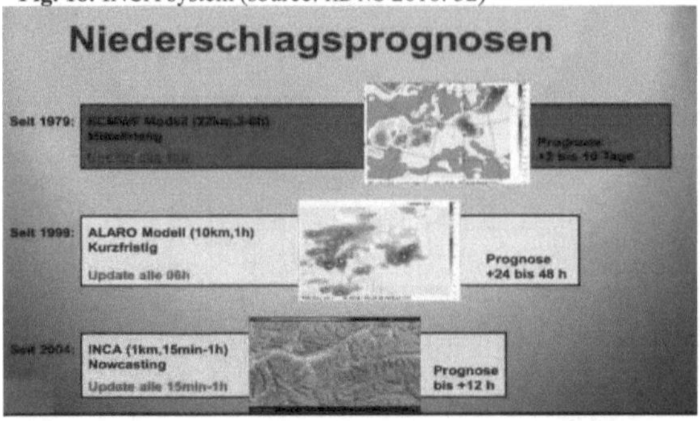

Fig. 19: Models for precipitation forecasts (source: HD NÖ 2016: 29)

As can be seen in Figures 20 and 21, in addition to the so-called main run, an ensemble forecast is also calculated to estimate the uncertainty in the meteorological forecasts. Ensemble forecasts have been used in meteorology for years to determine forecast uncertainties. An ensemble refers to the totality of numerous different forecast runs. Ensembles are usually generated by repeatedly starting the numerical weather model based on different initial conditions. The range of variation of the ensembles ultimately serves to describe the uncertainty of the main run (deterministic run) of the forecasts of atmospheric variables (e.g. precipitation) (cf. AMT DER OBERÖSTERREICHISCHEN LANDESREGIERUNG 2013: 2; cf. KOMMA et al. 2006: 279).

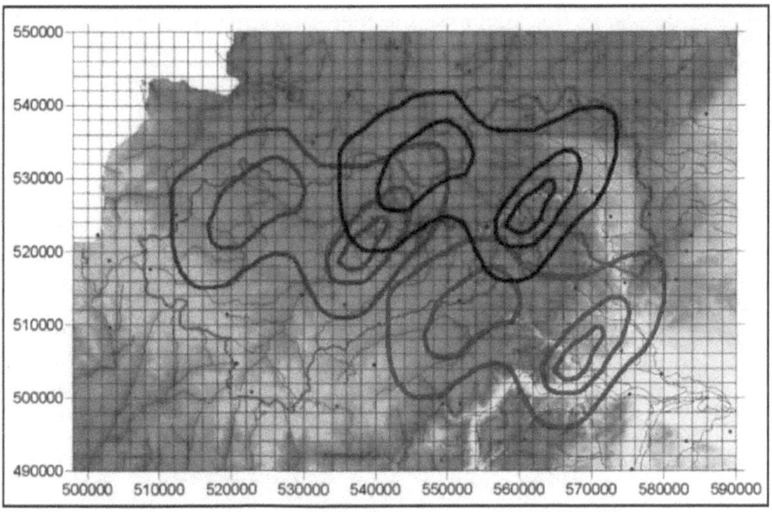

Fig. 20: Precipitation ensembles – spatial distribution (source: HDNÖ2016: 36)

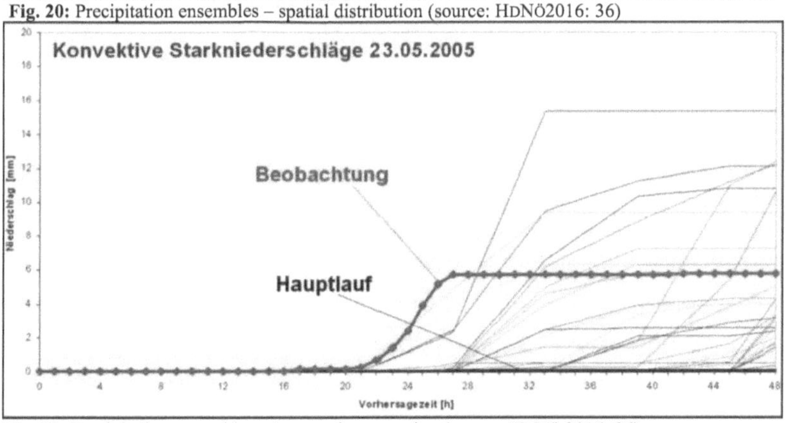

Fig. 21: Precipitation ensembles – temporal progression (source: HDNÖ 2016: 36)

6.2.2Focus 2 – Hydrological data

In step two, hydrological data is entered into the forecasting system. The hydrological model calculates the resulting discharge based on the meteorological forecasts and the current conditions in the Danube catchment area. The locations of the measurement points for the discharge are shown in Figure 22. The 92,000 ^{km2} catchment area of the Danube is divided into 90 sub-areas for the purpose of this calculation. The hydrological model calculates the discharge for each of these sub-areas, which is determined, among other things, by the precipitation that has fallen (cf. OFFICE OF THE UPPER AUSTRIAN PROVINCIAL GOVERNMENT 2013: 3).

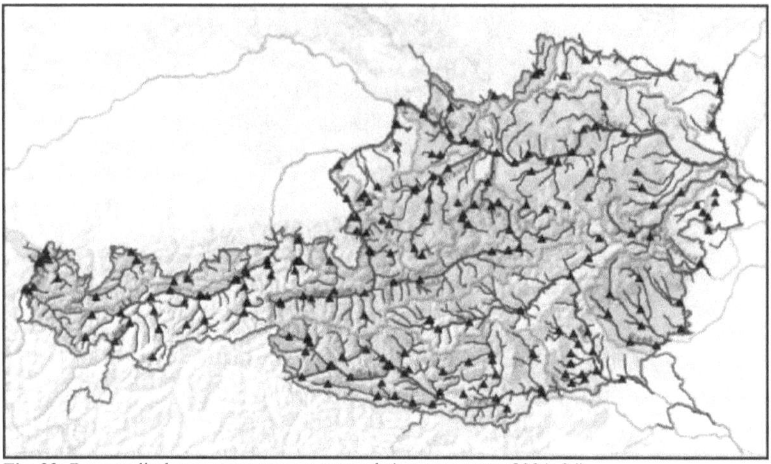

Fig. 22: Remote discharge measurement network (source: BMVIT 2004: 34)

> "As a result, the hydrological model provides the inflows expected for the next 48 hours for 36 inflow points to the Danube, whereby these inflow points include both significant tributaries such as the Bavarian Danube, Inn, Traun, Enns, etc., but also diffuse inflows from sub-areas." (OFFICE OF THE UPPER AUSTRIAN PROVINCIAL GOVERNMENT 2013: 3).

However, the public can only view the first 24 hours, as only reliable forecasts are published (cf. OFFICE OF THE UPPER AUSTRIAN PROVINCIAL GOVERNMENT 2013: 3).

6.2.3Focus 3 – Hydrodynamic data

The next step is to calculate the flow processes in the Danube and its forelands ( ). In collaboration with engineering firms, the University of Natural Resources and Applied Sciences and ZAMG developed a hydrodynamic model that maps the entire Danube between Passau and the

Austrian-Slovakian border, covering a length of approximately 350 km. This model simulates the behaviour of the ten Danube power plants as defined in their operating regulations, as well as flooding and runoff in the forelands (see AMT DER OBERÖSTERREICHISCHEN LANDESREGIERUNG 2013: 3f.).

6.2.4Focus 4 – Sequence control & visualisation

The final component in the forecasting system is the software for process control and visualisation, also known as the "forecast manager". As shown in Figure 23, this programme is used to coordinate all processes and to visualise input data and results (cf. SCIETEC 2006: 5).The forecast computer, shown in Figure 24, forms the core of the forecasting system. It is located in the computer centres or message centres of the hydrographic services of Lower and Upper Austria and functions in much the same way in terms of its basic design (HD NIEDERÖSTERREICH 2016: 22).

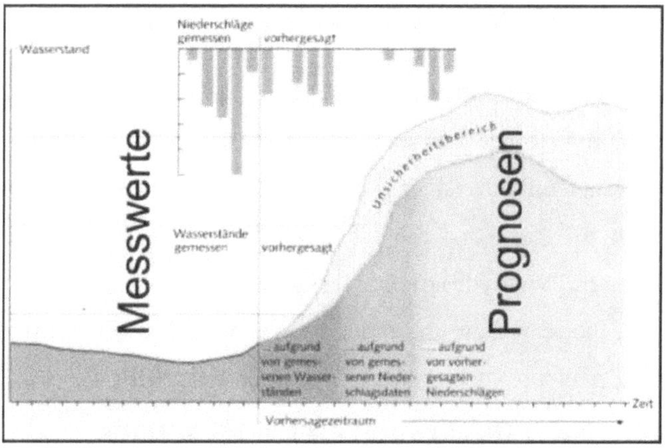

Fig. 23: Schematic diagram of the forecasting model (source: HD NÖ 2016: 22)

"In both federal states, calculations are carried out for the entire Danube with all its tributaries. Differences *arise from the fact that different areas have to be dealt with [...] and that the forecast computers in Upper Austria and Lower Austria are used for additional, different tasks [...]. The 'forecast computers' are redundant and designed as a subsystem consisting of several computers. The forecast computers run the hydrological and hydraulic models, the software for process control and visualisation, and the database in which all data relevant to the forecast is stored"* (SCIETEC 2006: 6).

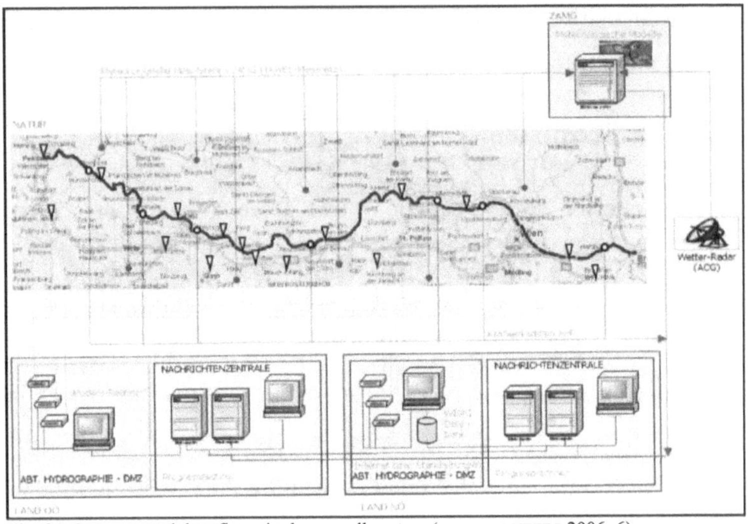

Fig. 24: Structure and data flows in the overall system (source: SCIETEC 2006: 6)

Although the systems in Lower Austria and Upper Austria operate independently of each other, they use the same database for their calculations. Corrections or other interventions carried out in one of the two federal states are automatically transferred to the system of the other federal state. This results in forecasts that are identical in content and easy to communicate between the two hydrographic services. Furthermore, if one of the systems fails ( ), the data from the other federal state is available (cf. OFFICE OF THE UPPER AUSTRIAN PROVINCIAL GOVERNMENT 2013: 4).

6.3 Difference between forecasting and early warning

Germany is already one step ahead in the field of flood forecasting. In 2000, a new forecasting system based on the LARSIM water balance model was put into operation in Austria's neighbouring country (see BREMICKER et al. 2006: 47).

"LARSIM uses extensive meteorological measurement and forecast data *to continuously calculate* all relevant hydrological processes, in particular current evaporation and the *soil water balance in the catchment areas"* (BREMICKER et al. 2006: 47).

Austria's models can only predict two days in advance. Nevertheless, Austria obtains data from the international
ECMWF model (see Chapter 7.2.1), which enables forecasts up to ten days in advance.

Flood forecasts are therefore divided into a reliable and an unreliable time range. The reliable time range is considered a reliable forecast and the unreliable time range is considered an estimate or early flood warning. The size of the catchment area of the gauge and the discharge conditions determine the length of the forecast classified as reliable. Since precipitation forecasts involve additional uncertainties, the forecast time for floods is significantly shorter than in the low and medium water areas

for floods is significantly shorter than in the low and medium water range (cf. BREMICKER et al. 2006: 47f.).

6.3.1 Flood forecasting

Flood forecasting is the provision of reliable information on expected short-term water level developments, which is usually updated hourly. The forecast horizon is relatively short at 48 hours in order to keep the differences between the predicted and actual water levels to within +/- 10 centimetres as far as possible. Based on flood forecasts, flood damage has been reduced and hazard prevention improved in recent years (cf. BREMICKER et al. 2006: 47).

"Flood forecasting is a reliable forecast for shorter time periods in accordance with the above accuracy criterion" (BREMICKER et al. 2006: 47).

6.3.2 Early flood warning

"Due to the relatively short time horizon of flood forecasts, the authorities responsible for protective measures and the citizens affected often need to receive information about the magnitude and timing of expected water level rises much earlier than has been the case to date" (BREMICKER et al. 2006: 46).

In Germany, a new forecasting system based on the LARSIM water balance model was introduced in 2000. Seven-day forecasts are possible with the conversion of the modelling technology to LARSIM. However, the aforementioned accuracy criterion of +/- 10 centimetres can only be achieved for a shorter time period (cf. BREMICKER et al. 2006: 47).

> "Flood early warning is a less reliable estimate that refers to a subsequent time period of up to seven days in total, depending on the model ( )" (BREMICKER et al. 2006: 47).

This allows a distinction to be made between the terms flood forecast and flood early warning (see BREMICKER et al. 2006: 47).

Figure 25 graphically illustrates and lists the key distinguishing features between flood forecasts and early flood warnings. Due to their longer lead time, early flood warnings are subject to greater uncertainty (cf. BREMICKER et al. 2006: 47).

	HW early warning	HW forecast
Publication	All year round, outside or in advance of floods	During a flood
Update frequency	once a day	hourly
Forecast time horizon	Up to 7 days	4 to 24 hours (depending on river basin)
Desired accuracy	Order of magnitude of possible water level rises (approx. +/-50 cm)	+/-10 centimetres
Possible use	Preliminary planning for Expected flooding	implementation of short-term Protective measures (mobile walls, sandbags, etc.)

Fig. 25: Distinguishing features between early flood warning and flood forecasting (source: BREMICKER et al. 2006: 47)

6.3.3Benefits of early flood warnings

The reliability of early flood warnings depends to a large extent on reliable weather forecasts. Due to the possibility that inaccurate early warnings may be issued, the authorities, industry and citizens affected are advised to obtain sufficient information before taking any action in the event of a flood warning. Early flood warnings should, among other things, contribute to a longer preparation time for flood response for municipalities, disaster control and industry (cf. BREMICKER et al. 2006: 49).

6.4 The European Flood Alert System (EFAS)

The European Flood Alert System (EFAS) is an EU project and a further development of the European Flood Forecasting System (EFFS) (see BUREK et al. 2011: 193).

The aim of EFAS is to improve disaster management in Europe by reducing the impact of floods through early warning. The existing flood forecasting models of the individual countries are not to be replaced; instead, EFAS is only intended to inform the national flood centres about possible flood developments. To this end, forecasts up to ten days in advance are to be used (cf. BUREK et al. 2011: 193f.).

The system is based on the GIS-based precipitation-runoff model LISFLOOD. LISFLOOD is a grid-based, physically based model consisting of a water balance module, a flood module and a module for modelling flood areas. Meteorological forecasts are also required in order to achieve a forecast period of up to ten days for flood forecasts. For this purpose, forecasts from the European Centre for Medium-Range Weather Forecasts (ECMWF), the German Weather Service (DWD) and COSMO-LEPS with a forecast time of four to ten days are used (cf. BUREK et al. 2011: 1934).

EFAS is a probabilistic flood early warning system and is therefore based on a bundle of meteorological forecasts, known as ensembles (cf. BUREK et al. 2011: 196).

> "Each of these meteorological forecasts represents the initial conditions for a LISFLOOD simulation and results in a possible future hydrological water status, so that ultimately a corresponding discharge forecast exists for each ensemble member. In contrast to deterministic forecasts, the uncertainties in the meteorological forecast are taken into account" (BUREK et al. 2011: 196).

However, this results in a probabilistic relationship rather than a clear statement. In order to be able to use the probabilistic statements as decision-making aids, the criterion of limit value exceedance was introduced, among other things. The limit value exceedance requires critical threshold values that categorise the hazard levels of the expected flood as shown in Figure 26. These threshold values are based on observed meteorological data (cf. BUREK et al. 2011: 196).

EFAS threshold value	Colour	Hazard description
S (extreme)		Threshold value corresponds to a simulated flood event with a return period >20 years Potentially extreme flood event expected
H (high)		Threshold value corresponds to a simulated flood event with a recurrence interval of >5 years and <20 years Potentially significant flood expected
M (medium)		Threshold value corresponds to a simulated flood event with a recurrence interval of >2 years and <5 years Water level expected to reach or slightly exceed maximum capacity. No significant damage expected in the event of flooding
L (low)		Threshold value corresponds to a simulated flood event with a return period of >1.5 years and <2 years Water level higher than normal, but no flooding expected

Fig. 26: EFAS threshold values (source: BUREK et al. 2011: 196)

In order to summarise the information provided by the system in a way that is easy for national hydrological services to understand and use, particular emphasis is placed on developing methods for visualising the forecasts. This is especially important because EFAS results are disseminated throughout Europe in English, but English is often not the native language of the recipients and could therefore be misunderstood. ly, EFAS forecast products focus on presenting threshold exceedances in the form of maps, diagrams and tables (see BU- REK et al. 2011: 197).

By 2006, EFAS had been implemented for the entire Danube catchment area. Austria has been a member of the EFAS group since then. The Danube catchment area is modelled on a 1x1 km grid. In order to carry out the necessary data collection, calibration and validation of the model, delegates from the countries in this catchment area were assigned to the EFAS Joint Research Centre for two to three years (cf. BUREK et al. 2011: 194).

6.5 Flow chart of a Danube flood early warning system

This warning is intended to alert all authorities, agencies, emergency

This warning is intended to reach all authorities, agencies, emergency services and residents in the areas at risk of an impending Danube flood event. To this end, the province of Lower Austria has designed a flow chart of the reporting channels, as shown in Figure 27. This flow chart is divided into a warning and an alarm. Chapter 7.5.4 describes the step-by-step warning measures using the example of Krems. The alarm measures are implemented by prioritising five set-up and dismantling phases of mobile flood protection. This is explained in Chapter 8.2.1 using the municipality of Weißenkirchen as an example (OFFICE OF THE LOWER AUSTRIAN PROVINCIAL GOVERNMENT 2016: 3).

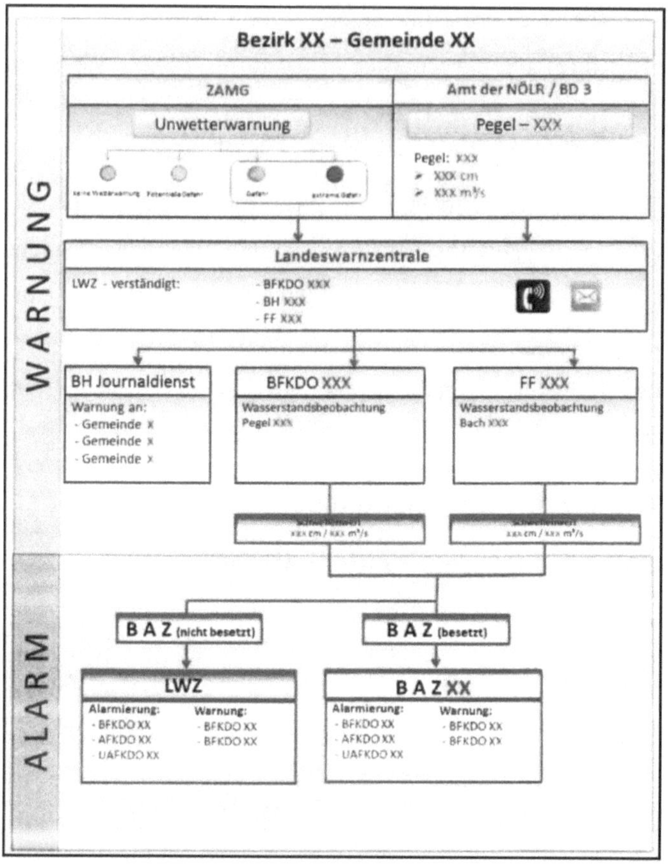

Fig. 27: Flood early warning reporting channels (source: OFFICE OF THE LOWER AUSTRIAN PROVINCIAL GOVERNMENT 2016: 3)

6.5.1 Triggering the warning

The State Warning Centre (Fire Brigade and Civil Defence Department) is automatically informed by the Hydrology and Geoinformation Department and additionally by telephone and e-mail when a previously defined water level is reached or exceeded. After consulting with the Hydrology and Geoinformation Department, the Provincial Warning Centre notifies the district administration during office hours and, outside office hours, the district control centre, the responsible police station (PI) and the local fire brigade (see KREMS DISTRICT ADMINISTRATION 2014: 4).

6.5.2 Water level observations

The water level is checked once a day on working days by the relevant district authority on the Internet or from a tape recording. Once the warning has been triggered, the water level is checked hourly by the district authority via the Internet, in consultation with the hydrology department or from the tape recording (see KREMS DISTRICT AUTHORITY 2014: 5).

6.5.3 Forwarding the warning

As a rule, the warning message must be sent to the specified recipients by e-mail, fax, telephone or radio, possibly also with a forecast. However, when transmitting by email and fax, it must be ensured that receipt of the messages is confirmed. If the flood event occurs unexpectedly and an extremely rapid rise is to be expected, the population must be warned by loudspeaker, leaflet or

personal communication. In this case, the authorities may also, on the instructions of the head of operations (district administrator), trigger the alarm by means of a siren signal. When this signal is used, a distinction is made between "danger imminent" and "no danger imminent". If there is no imminent danger, the state warning centre is notified with a request to issue a civil defence warning for the area "xy". The state warning centre must clarify the radio announcement, which must be linked to the civil defence signal "warning", with the ORF as quickly as possible and issue the civil defence warning for the endangered area. , the provincial warning centre, the district administration and the endangered area(s) must confirm that the civil defence warning has been issued and that it is functioning properly (see KREMS DISTRICT ADMINISTRATION 2014: 6).

In the event that no connection can be established with the provincial warning centre or the ORF cannot be notified for the radio announcement and

If there is imminent danger to the population due to the event, the siren signal "Civil Defence Warning" must be sounded as a warning on the orders of the authorities (district administration, municipality, etc.) from the nearest possible trigger point (state warning centre, regional or district alarm centre, local fire brigades). In the event of locally triggered civil defence signals without prior involvement of the district administrative authority and state warning centre, the nearest authority must be informed as soon as possible (see DISTRICT ADMINISTRATION KREMS 2014: 6f.).

7.4.1 Measures in the event of early flood warnings

During office hours, the district administrator and the disaster control officer must be notified immediately. Outside office hours, if the alarm is triggered via the district control centre (PI), the latter shall immediately notify the district administration's duty officer and the disaster control officer. Upon notification by the district control centre (PI) or directly by the state warning centre, the duty officer shall notify the district administrator and contact the disaster control officer. The duty officer and the disaster control officer shall then proceed to the office and carry out the tasks in accordance with the alarm plan. Until the district administration office is staffed, the district control centre (PI) handles any further reports received regarding concerning an approaching Danube flood (cf. KREMS DISTRICT ADMINISTRATION 2014: 8).

With regard to the city of Krems an der Donau, the following measures are taken in the event of an approaching flood (cf. KREMS DISTRICT ADMINISTRATION 2014: 8ff.).

When the warning level "Kienstock" of a water level of 6.15 m is reached, the Krems district administration office is manned by disaster control officers, the duty officer(s) and other staff as required (IT, etc.). Outside office hours, a telephone switch to the district control centre (PI) must be carried out (see KREMS DISTRICT ADMINISTRATION 2014: 8).

When the warning level "Kienstock" is reached at a water level of 6.50 m, the municipalities must issue a warning to the population (see KREMS DISTRICT ADMINISTRATION 2014: 11).

When the warning level "Kienstock" is reached at a water level of 7.02 m, the large overflow section is activated. This serves as a relief facility, which releases part of the discharge into the hinterland via overflow sections when water levels are high (cf. BMLFUW 2006: 6f.). From this level, regular dam patrols must be carried out (see KREMS DISTRICT AUTHORITY 2014: 11).

When the warning level "Kienstock" is reached at a water level of 7.15 m, all affected roads are closed and shipping is suspended (see BEZIRKSHAUPTMANNSCHAFT KREMS 2014: 12).

CHAPTER 7

Flood protection

For economic and strategic reasons, many towns and municipalities were built on rivers in the past. However, the areas used were preferably located on elevated banks outside the floodplains and were therefore protected from flooding in most cases. Over time, economic development increased, leading to a concentration of people and property in urban areas and intensive use of previously unused land near watercourses. People and property are particularly at risk from flooding in these areas. Nowadays, it is only in exceptional cases that land is taken out of use or even resettled. Flood protection measures such as dams or retention basins require a large amount of space and cannot be implemented in urban areas due to the often limited space available.

Flood damage can be reduced with the help of structural measures, targeted preparation of emergency services and optimal coordination of all protective measures in the event of flooding (cf. PATT 2001: 225).

In order to establish the connection to Section 5.2, the following subsection refers to mobile flood protection. This flood protection measure in particular only works with a hydrological flood forecast and an advance warning time that allows it to be set up for every potential damaging event. In the Wachau region in particular, mobile flood protection is the only option for effective protection due to space constraints (BMLFUW 2002: 33).

7.1 Mobile flood protection

> "Flood protection structures that are only erected during a flood event supplement dykes and protective walls in areas where permanent structures are not possible or desirable" (PATT 2001: 284).

Once the alarm threshold has been exceeded, the mobile protection systems must be transported from the storage facility to the site of deployment and then erected. They are dismantled and returned to the storage facility immediately after the flood event. As can be seen in Figure 28, mobile flood protection walls consist of vertical supports, support or retaining structures for these supports and wall elements located between them. Intermediate supports are also required for longer walls (cf. PATT 2001: 284).

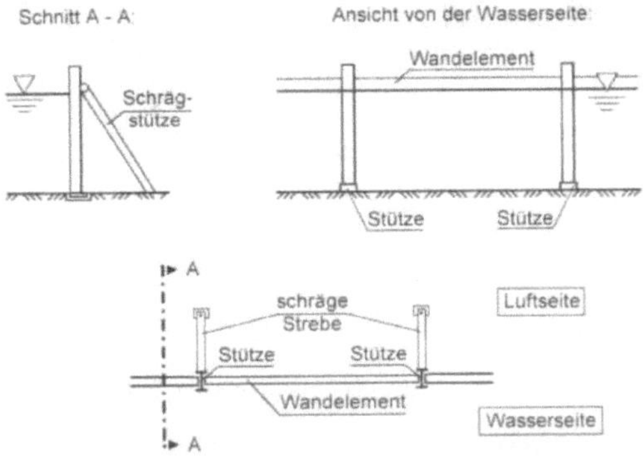

Fig. 28: Mobile flood protection wall (source: PATT 2001: 287)

7.2 Mobile flood protection in the cadastral municipality of Weißenkirchen/Wachau

In 2010, the construction of a mobile flood protection system in the cadastral municipality of Weißenkirchen in the Wachau region, which also includes the municipalities of Wösendorf and Joching, was completed at a cost of around 27 million euros. This aluminium dam system is an EKO system from Kossbiel ( ) and is installed on a specially designed flood protection wall, as shown in Figure 29 (DENK 2016).

Fig. 29: Flood protection wall in Weißenkirchen (source: HASENHÜTL)

The protection system can also be adjusted in stages to the expected flood level and is insensitive to dimensional inaccuracies and temperature fluctuations. To prevent water leakage, sealing systems made of stainless steel are integrated between the dam beams, uprights and floor rails. The dimensions and material of the seals are the same in all areas to ensure quick replacement after mechanical damage (DENK 2016).

The individual beam elements, each 2.57 m long and weighing around 25 kg, are inserted into pre-mounted steel supports as shown in Figure 30. Thanks to a carefully developed installation plan and extensive standardisation, the system ensures virtually error-free installation of the system beams (DENK 2016).

The mobile system consists of 1,572 supports, which are shown in Figure

31. However, only four of the 1,572 supports are of a special format and 60 are used as back supports. The special supports are placed at corner points where a triple continuation of walls is required (DENK 2016).

Fig. 31: Steel supports (source: HASENHÜTL)

As can be seen in Figure 32, 7689 dam beams are used, 957 of which have a floor seal (DENK 2016).

Fig. 32: Dammbalken (source: HASENHÜTL)

The low weight of the beams allows the population to participate in the construction of the mobile protective wall. In addition, all dam beams are designed so that they can only be installed in the correct position. This also applies to the supports, as the base plates for support mounting and support bolting do not allow them to be inserted at an angle. Assembly and dismantling are generally carried out by municipal employees and the volunteer fire brigades of Weißenkirchen, Wösendorf and Joching according to a pre-established operational plan. As shown in Figure 33, training sessions are held every six months with the Austrian Armed Forces, who provide support in the event of flooding (DENK 2016).

Fig. 33: Austrian Armed Forces training on the Weißenkirchen flood protection system (source: HASENHÜTL)

The mobile flood protection system built in the cadastral municipality of Weißenkirchen an der Donau, with a total length of 2.9 km stretching from Wösendorf to Weißenkirchen, is Austria's largest flood protection system ( ). The wall height that can be achieved is 3.5 m and the mobile protection area covers 6300 $^{m^2}$(DENK 2016).

As shown in Figure 34, a storage hall had to be built to store the mobile elements, which also provides modern facilities for the management team and operations control in crisis situations (DENK 2016).

Fig. 34: Flood protection hall in Weißenkirchen (source: HASENHÜTL)

It is not only the masses of water but above all the mud deposits left behind by these Danube floods that cause major damage. As this aluminium dam beam system, as shown in Figure 35, consists entirely of cavities that can fill with water and thus provide their full protective effect, mud also accumulates in the dam beams (DENK 2016).

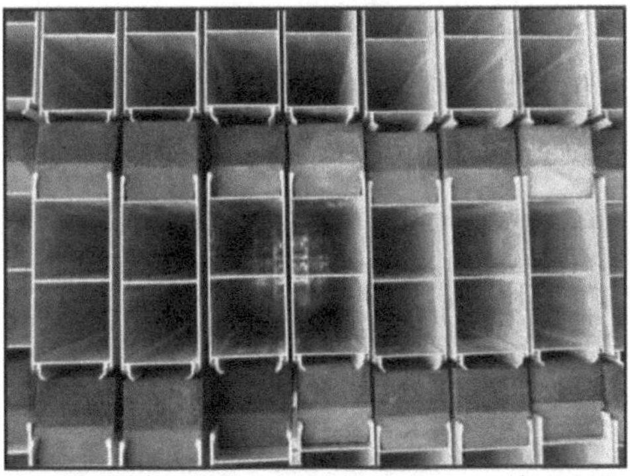

Fig. 35: Cavities in the dam beams (source: HASENHÜTL)

As this sludge is particularly aggressive and becomes concrete-like when dry, HTL Hollabrunn developed and built a specially designed dam beam washing system, as shown in Figure 36 (DENK 2016).

Fig. 36: Washing system for the dam beams for the flood protection system (source: HASENHÜTL)

7.2.1 Construction of the mobile flood protection system in Weißenkirchen

To set up the mobile elements, a phase plan was drawn up by the municipality of Weißenkirchen based on the "Kienstock" water level as part of a special alarm plan, which is shown in Table 4. This phase plan is intended to explain which measures relating to flood protection setup/teardown should be taken when the water levels sent by the Hydrology Department are reached. Phase 0 is still in the early warning stage. The alarm, as shown in Chapter 7.5, begins with Phase 1 (WEIßENKIRCHEN ALARM PLAN 2016: 10).

Phase	Water level Kienstock	Protective function
Phase	Early warning stage	
Phase 1	7.50	Protection up to approx. 9.00 m
Phase 2	8.20	Protection up to approx. 9.50 m
Phase	8.60	Protection up to approx. 10.00 m
Phase 4	9.00 m	Protection up to approx. 10.50 m
Phase 5	9.90	Protection up to approx. 11.00 m

Table 4: Phase plan (data source: WEIßENKIRCHEN 2016: 10; own representation)

Based on forecasts, it is also possible to start setting up an emergency phase earlier. According to the manufacturer, the set-up time between phases is four hours. These times were confirmed during the emergency drills (WEIßENKIRCHEN ALARM PLAN 2016: 10).

As soon as the "Kienstock" level reaches a height of 10.30 m, evacuation must begin immediately (WEIßENKIRCHEN ALARM PLAN 2016: 10).

The following tables (Tables 5, 6, 7, 8 and 9) show the individual phases in detail and list the materials required for each phase (WEIßENKIRCHEN 2016).

Phase		
Close until	Floor-level supports	64
Top edge of wall	Dam beams with floor seals	99
(8.96 m)	Dam beams	264
Kienstock	Special supports	4

Table 5: Phase 1 (data source: WEIßENKIRCHEN 2016: 10; own representation)

Phase		
2.50 m	2.50 m and 2.25 m supports	464
	Dam beams with floor seals	462

Supports and 2 stop logs	Dam beams	616

Table 6: Phase 2 (data source: WEIßENKIRCHEN 2016: 10; own representation)

Phase		
2.00 m supports and 2 dam beams +2 dam beams on top of the existing ones (4 dam beams in total)	2.00 m supports	168
	Dam beams with floor seals	165
	Dam beams	1232

Table 7: Phase 3 (data source: WEIßENKIRCHEN 2016: 10; own representation)

Phase		
1.50 m supports and 3 dam beams dividing the supports +3 dam beams on top of the existing ones (7 dam beams in total)	1.50 m supports	192
	Dam beams with floor seals	231
	Dam beams	2376
	dividing supports	620
	Back supports	60

Table 8: Phase 4 (data source: WEIßENKIRCHEN 2016: 10; own representation)

Phase 5		
3 dam beams (10 in total)	Dam beams	2244

Table 9: Phase 5 (data source: WEIßENKIRCHEN 2016: 10; own representation)

CHAPTER

Empirical

This part of the paper aims to answer the key research question: "What are the functions of flood forecasting systems and early warning systems?" In addition, the results of the qualitative study will be evaluated and presented.

CHAPTER

Method

In order to examine the research questions and prove the hypotheses, two working methods were used in this thesis. On the one hand, literature research was used, particularly in the first theoretical part of the thesis, to create a theoretical basis. The first two research questions and their hypothetical evidence were examined and verified in the theoretical part of the thesis. These are the research question "How much does climate change contribute to the increased occurrence of floods?" and another research question "What factors are necessary for a functioning mobile flood protection system?" On the other hand, a qualitative method of guided interviews, which is reflected in the second part of the thesis, was used to examine the third research question, "What is the function of flood prediction systems and early warning systems?" and to test the corresponding hypothesis.

In the first theoretical part of the thesis, a comprehensive overview of the topic was provided by reviewing numerous written works. The first step was to obtain relevant literature from university libraries, which, however, yielded only limited results. As a result, it became necessary to organise published and unpublished publications and verbal information from the relevant institutions. An expert from Austro-Control assisted me with regard to "weather". I was able to obtain information on the distinction between the two terms "flood forecasting" and "flood early warning", which are difficult to separate, from an employee of the Federal Ministry of Agriculture, Forestry, Environment and Water Management. In addition, a field day was organised in Weißenkirchen in the Wachau region to gain an insight into the mobile flood protection system.

In order to carry out further practical data collection, the author of this thesis decided that the question regarding flood forecasting and early flood warning could best be answered in the form of a qualitative study. For this reason, four interviews were conducted with experts from four different institutions on the subject of flood forecasting. The interviews lasted an average of one hour. The interview guide can be found in chapter 10.3 and the transcripts are available in the appendix. For data protection reasons, the audio files are available from the author of this thesis.

9.1 Survey method

The survey method is a guided interview with experts. The interviewees include a number of experts from a wide range of institutions, such as the Lower Austrian and Upper Austrian provincial governments, the Lower Austrian Provincial Warning Centre and the Lower Austrian Civil Protection Association.

The expert interview is a special form of guided interview. In this type of interview, the interviewees are interesting as experts in a specific field. Who is considered an expert depends largely on the subject of the study and varies due to the theoretical and analytical research approach (cf. FLICK 2011: 214). In order to ensure better comparability between several interview partners, the questions were structured. This structuring of the questions and their scope was therefore also carried out by the interviewer. The mostly open- ly formulated questions that the guided interview entails are a central feature of this form of survey. The interviewee should be able to answer the questions as freely as possible. Each question asked should encourage the interviewee to talk, but also give them the opportunity to express their thoughts and opinions. Since the questions do not have to be asked in the order specified in the guide, the interviewee can decide for themselves when they want to answer which question. However, the interviewer must ensure that all questions in the guide are asked or indirectly answered during the interview. Some questions may be omitted if they are already answered in the flow of the interviewee's narrative. This gives the interviewer the opportunity to steer and shape the interview individually (cf. FLICK 2011: 222f.).

One advantage of this methodology is the information gathering, as it is quick and comprehensive. In addition, all information obtained can be recorded for a long time via video and audio recordings as well as transcription files. Results and procedures can be traced at any time thanks to the aforementioned types of documentation (cf. FLICK 2011: 224).

As all interviews took place in the offices of the interviewees, a pleasant atmosphere similar to that of an everyday conversation could be ensured. The manner in which the questions were asked was also essential. They were short, simple and easy to understand, and complicated, exaggerated technical language was avoided (cf. REINDERS 2005: 96).

9.2 Evaluation method

The transcribed interviews were analysed to identify commonalities and comparable content. According to FLICK (2002: 279), qualitative content analysis is the most classic method for analysing interview data. A key feature of this method is the use of categories and the reduction of the available material. The interview guide was used to categorise thematically similar areas. This schematisation can be found in chapter 10.4 Category system. Statements made by the interviewees on the individual categories were summarised separately. The complexity of the results was reduced by summarising the statements. An interpretation of the information results can be found after each summary. In order to answer the research question, the empirical results were confronted with the theoretical context in the interpretation (GLÄSER and LAUDEL 2010).

9.3 Interview guide

1. Training and qualifications

a) What training did you have to undergo to perform your duties?
b) What is your professional background?
c) How did you ultimately end up at this institution?

2. Institution and field of activity

a) Could you describe how your company is organised?
b) What are your specific responsibilities, or do you also perform other tasks?
c) To what extent do you cooperate with colleagues or an inter y team in your daily work?

3. Information gathering and forecasting systems

a) What tools and resources are available to you to perform your current tasks?
b) Which sources of information do you use on a daily basis?
c) Which of these sources are most meaningful and actionable for you?
d) How do you manage the large amount of information available to you?
e) How do you obtain current water levels and water level forecasts?
f) Which measuring stations do you use for water level forecasts? (Precipitation measuring stations)
g) Where are these measuring stations located?
h) How many of these are currently in use?
i) What processes and procedures support your work?
j) Are there different water levels?
k) How are the warning levels divided?
l) Are there measures in place when an early warning is triggered? (Closing

locks, sandbags, alerting emergency services)

m) How do you set up hydrographic measuring stations and how are they maintained?

n) How maintenance-intensive are these services?

4. Use of data

a) What purposes does your work serve?
b) Who has access to your results?
c) Who are the users of the results?
d) Is the data obtained sold?
e) How do you process the measurement data?
f) Is it processed electronically?
g) How are the data statistically evaluated and regionalised?
h) How are hydrological reports prepared?
i) What is the difference between hydrographic reporting and a news service?
j) What is the difference between a flood forecast and an early flood warning?

5. Decisions and financing

a) What are the potential consequences of the decisions you make? (e.g. evacuation)
b) What measures do you see to improve the effectiveness and efficiency of your services/forecasts?
c) Do you tend to make short-term or long-term decisions?
d) How do you finance your services?

The same guideline was used for all four interviewees. However, for reasons of competence, not all interviewees were able to answer all questions in full. Due to the specific topics of interest to the experts, the order of the guideline was not always followed. However, thanks to the pleasant atmosphere of the interviews and the open-ended questions, this was not perceived as a disruptive factor.

9.4 Category system

The author used the following categories to structure the interview guide.

- Institution and field of activity
- Information gathering and forecasting systems
- Use of data yields
- Impact of decisions

9.5 Introduction of the interviewees

The interviewees were selected to ensure that experts from various sub-areas of flood forecasting were represented. According to GLÄSER and LAUDEL (2010), each expert has a unique perspective on the subject matter under investigation due to their individual position and personal observations.

The following presentation of the interview partners is based on the first question in the interview guide. The experts describe their connection to flood forecasting, thereby demonstrating their suitability as competent discussion partners for this thesis. The experts are listed in the order in which the interviews were conducted, and this order is also used in the evaluation of the categories.

Expert 1

E1 comes from military service. He served in the Austrian Armed Forces for 16 years and is a trained non-commissioned officer. As the activities of the Civil Protection Association are very diverse, this interview focuses on his experience with this organisation.
On the one hand, E1 gained his experience during his military training, as he had to complete all relevant specialist training, including leadership training.

On the other hand, he can also contribute personal experience, as his home has been flooded several times. Via a detour into voluntary work, which saw him complete all the courses offered by the Civil Defence Association over a period of ten years and take on various areas of responsibility, his path ultimately led him from the Austrian Armed Forces to the Lower Austrian Civil Defence Association, where he now holds the position of Regional Manager.

Expert 2

E2 studied at the University of Vienna and is a geologist specialising in hydrogeology. After successfully completing his studies, he worked for a civil engineering firm for three years. When this company was dissolved, he set up his own business for another three years. In 2000, he was appointed head of the water level information service in the Lower Austrian provincial government's Department of Hydrology and Geoinformation.

E3 studied at the University of Natural Resources and Life Sciences and wrote his thesis on disaster control. Through extensive research and investigation in this field and many years of voluntary work in an emergency services organisation, he was able to gain a wealth of experience and acquire relevant qualifications.
E3 worked in research at the University of Natural Resources and Life Sciences until the end of 2002 and began working for the state government in the fire and civil defence department in 2003.

Expert 4

E4 also studied at the University of Natural Resources and Life Sciences, specialising in hydraulic engineering and hydrology. After graduating, he worked for VIADONAU for nine years as head of hydrology and flood services. He then moved to the Upper Austrian provincial government, where he worked in the surface water and hydrography department.

CHAPTER 10

Evaluation of the interviews according to the category system

10.1 Institution and field of activity

Expert

E1 explains that the Lower Austrian Civil Protection Association is an association with its own statutes, consisting of eleven full-time and around 1,200 volunteer staff. This makes it the largest civil protection association in Lower Austria. The tasks of the Civil Protection Association include processing state orders and training the population in areas such as flood disaster management. He also mentions: "There are orders that require us to carry out activities such as training in crisis and disaster management at the municipal level, i.e. for municipal representatives and mayors. This means that the state government has commissioned us *to provide this* training and *support for training*" (E1: lines 20-24). The Lower Austrian Civil Protection Association and the state of Lower Austria therefore work closely together. On the one hand, the association is financed by the state, and on the other hand, due to their proximity – they are located in the same building – joint projects are carried out on a regular basis. E1 currently holds the position of interim state manager in this association until the next state executive committee meeting on 25 February. His activities include giving lectures and moderating discussions on topics such as radiation protection, flooding, mudslides and much more (E1: lines 2–36).

Expert 2

E2 states that the Lower Austrian state government is divided into many different departments ( ). E2 works in the Hydrology and Geoinformation Department, which is in turn divided into several divisions. As head of the

Water Level Information Service Division, his main responsibilities include preparing flood forecasts. He also mentions: "My job is therefore mainly the organisational and technical management and maintenance of this operation, so that in the event of flooding, the people who are actually providing the flood service have the appropriate infrastructure and are able to use it, which means that training is also part of my *job"* (E2: lines 12–16).
However, in order to be able to make flood forecasts, cooperation with other institutions is necessary. E2 therefore distinguishes between technical and informal cooperation. Technical cooperation takes place due to the necessary collaboration with the hydrology departments of neighbouring countries, the Czech Republic and Slovakia, as well as with the federal states of Upper Austria, Salzburg and Styria. We cooperate with the state warning centre on an informal basis. Nevertheless, we publish all data on the Internet and on teletext, among other places, and thus cooperate with the entire affected population (E2: line 3 - 27).

Expert 3

E3 explains that the state of Lower Austria has around 70 departments, which are in turn organised into groups. E3 works in one of these specialist departments, namely the Department of Fire and Civil Protection. This department is responsible for disaster control and its actions are based, among other things, on the Lower Austrian Disaster Control Act. It is also divided into three areas. The first area is disaster control planning and organisation, the second is warning and alarm system technology, and the third area is the state warning centre. E3 is the deputy head of the fire brigade and civil protection department. This means that he is also responsible for ensuring that staff work runs smoothly in the event of a disaster and that all necessary plans are in place. As a future task, E3 would also like to devote himself to the creation of new and better structured special alarm plans. E3 is currently working on a draft in this regard. These

special alarm plans should be individually editable for all municipalities in Austria, thus encouraging internal adaptation within the municipalities. E3 emphasises: "This means that the majority of my work involves standing in the classroom, teaching people, training people, organising exercises ranging from small-scale exercises at municipal or district level to large-scale regional exercises or cross-border exercises with neighbouring countries" (E3: lines 49-52).
As a sideline, he teaches at the University of Vienna's Institute of Geography and Regional Research. The Fire and Civil Defence Department is therefore a hub for all

disaster scenarios. This enables extensive cooperation with numerous specialists, including hydrologists, chemists and radiation experts (E3: lines 1–71).

Expert 4

E4 explains that the relevant hydrographic service in Upper Austria consists of several departments. In particular, there is a surface water department, which includes the flood service. E4 works in the environment department and supports the flood service in the relevant project work. With regard to cooperation with other institutions, E4 mentions that annual meetings are held with Germany to exchange experiences (E4: lines 1–21).

10.1.1 Summary

The interviewees come from a wide variety of organisations and therefore hold different positions with regard to flood forecasting. It is worth noting that all interviewees are either employed by the state itself or by organisations that are in turn financed by the state. It can therefore be concluded that the state is responsible for all research, training, development and prevention measures. All interviewees hold management positions and are responsible for different tasks in the event of a flood. In order to raise public awareness and thus improve the systems, ongoing exchanges of experience and cooperation with neighbouring countries as well as other organisations and institutions in Austria are necessary. All interviewees confirmed that this cooperation between the individual institutions works smoothly.

10.1.2 Interpretation

Flood forecasts are specific to each state and are anchored in different departments of the respective state governments. Due to the many different responsibilities and areas of responsibility before, during and after a flood event, it is necessary to act with special alarm plans, as is already partially practised in Lower Austria. These special alarm plans set out step by step how each municipality is to proceed individually in the event of a flood and which hierarchical steps must be followed in communication. Otherwise, smooth operations, which are essential for all parties involved, would be impossible. Since flood forecasts made by the state's hydrologists using state-of-the-art technology are the most important factor in minimising damage during a flood , training for the population and preventive measures are also essential. These are carried out by the state's fire and civil defence department on the one hand, and by the civil defence association on behalf of the state on the other. However, to avoid inconsistencies and possible overlaps, intensive coordination and cooperation between the state and the civil defence association is necessary. In Lower Austria, this cooperation between the two organisations works extremely well thanks to their shared infrastructure.

10.2 Information gathering and forecasting systems

Expert

E1 explains that the civil protection association obtains its information on flooding from various systems. The association receives the results data generated by the flood forecasting models from the Hydrology and Geoinformation Department of the Lower Austrian provincial government. The task of the civil protection association is to make the population aware of these facts. The association then motivates and trains people on where and how to access this flood information. The civil protection association primarily uses the "NÖ Atlas" to educate the population. This is a simple user programme that shows all the measuring stations on the Danube that transmit measurement data electronically to this "atlas". In the event of an emergency, citizens can use this programme to raise their awareness and determine whether their homes are located in the danger zone and at what water level they can expect water to enter their homes. In addition, the affected population is also warned by civil defence alarms in the form of siren signals. The task of the civil protection association is to train the population so that everyone can make a targeted contribution in the event of a disaster. With the help of published, municipality-specific special alarm plans, the civil protection association aims to raise awareness among the population and make them aware of their obligation to protect themselves. These special alarm plans also contain information tailored to the respective municipality.

Water level (warning) stages. These water level (warning) stages are directly linked to any countermeasures for flood control. The civil protection association tries to motivate and train the population in this regard so that they can cooperate with the emergency services and local authorities in the event of a disaster and not disrupt flood control operations through ill-considered actions (E1: lines 57–124).

Expert 2

E2 points out that the Department of Hydrology and Geoinformation has

many different sources of information at its disposal for preparing flood forecasts. E2 begins by mentioning that the province of Lower Austria operates its own precipitation measuring stations for flood forecasting, among other things. As these precipitation measuring stations can only provide location-specific data, additional data must be incorporated into the overall forecast. The province of Lower Austria obtains this additional data from the Central Institute for Meteorology and Geodynamics (ZAMG). The ZAMG uses the precipitation measurement data from the state, purchased radar data from Austro-Control, and its own and additional international precipitation models to create a "spatial precipitation distribution" optimised for Austria. The "spatial precipitation distribution forecast" is continuously adjusted to reflect changes.
E2 explains: "This is clear in the case of a thunderstorm, where I have a thunderstorm tower that can either move or *remain* in place" (E2: lines 104–109).

These precipitation cells are then assigned vectors that represent the direction in which the cells are moving. For a more extensive forecast into the future, the Europe-wide, large-scale weather model (ECMWF) and the small-scale weather model (ALADIN) developed by ZAMG are used. The large-scale weather model provides the input data for the small-scale weather model.
E2 explains: "There are precipitation situations where large-scale forecasts are simply better than small-scale ones *[...]"* (E2: 118 - 120).
After a 72-hour forecast, the ZAMG weather model (ALADIN) also reaches its limits. For further forecasts, only the ECMWF weather model is available, which calculates up to ten days into the future. In order to combine all weather models, ZAMG developed the "INCA" system.
Finally, the ZAMG provides a precipitation and temperature forecast, which is sent every 15 minutes to the user as central input for a flood forecast. This

forecast includes so-called "ensembles", which represent all possible precipitation probabilities.

The province of Lower Austria also obtains all fully evaluated flood forecasts from the Czech Republic, Upper Austria and Bavaria, which are incorporated into the flood forecast model as additional input data. The locations of the measuring stations operated by the province of Lower Austria for precipitation and water levels are clearly shown in the "NÖ Atlas" or on the

E2 mentions that the graph on the website contains both the current values as a hydrograph, i.e. a graphical representation of measured values in chronological order, and a deterministic forecast and extreme values, which are indicated by the 10% and 90% percentile of the precipitation forecast. E2 states that there is no hydrological uncertainty. E2 also explains that all measuring stations are always in active operation and are subject to intensive maintenance. A maintenance team remedies the causes of measurement errors, which can be caused by frost in winter and mud after flooding, among other things. E2 also has some information to share about the complexity of the water level measuring stations used to determine the water level. In principle, such a station consists of a built-in tape recorder, which documents the measured data in a water level measuring station, and a remote data transmission system. A built-in pressure probe measures the water pressure above the sensor, which then reports the water level. In addition, the province of Lower Austria employs so-called measuring point observers, who are employed on a part-time basis and are responsible for manually reading the water level. This allows the electronic data to be compared with the manual data in the "Hydrography and Geoinformation" department, enabling measurement errors in the electronic devices to be identified and the data corrected. On the other hand, a distinction can also be made between a relative, electronic measurement and an absolute, manual measurement. E2 also explains that these water level values are subsequently incorporated into a hydrological classification.

"We have mean water levels, annual floods, [...] five-year floods and 30-year floods" (E2: lines 218 - 220).

An "alarm manager", which is available as a software solution from the Hydrology and Geoinformation Department, refers to this hydrological classification and sends automated messages to the state warning centre. Countermeasures that must be taken when water levels are exceeded are then implemented either by the state warning centre or by the municipalities themselves (E2: lines 65–238).

Expert 3

E3 explains that the Fire and Civil Protection Department of the Lower Austrian Provincial Government also has numerous tools at its disposal for gathering information. E3 highlights the Alarm Manager, which two colleagues from the Provincial Warning Centre, which covers part of this department, have been trained to use. This "Alarm Manager" was developed by the "Hydrology and Geoinformation" department and is also owned by them. The "Alarm Manager" has now taken on a central role in transmitting warnings from the hydrological department to the state warning centre. The main task of the "Fire Brigade and Civil Defence" department, and thus of the State Warning Centre, is to pass on flood warnings. To this end, the employees of this department have access to telephones, computers, radio and a pager network operated by the State of Lower Austria. In cooperation with the ZAMG and the Hydrology and Geoinformation Department, the Fire and Civil Protection Department prepares a written forecast, which is sent out by the State Warning Centre via a defined, large distribution list. This distribution list includes all affected district authorities, municipalities, emergency services, infrastructure operators, such as hospitals, etc.

E3 adds: "As the event approaches, it is quite possible that something will be sent out twice a day. These are texts, forms, , which are simply inserted in a standardised format. We then use the forms when it is time *to issue* a warning″ (E3: lines 127–130).

The flood forecasts from the hydrology department and the precipitation forecasts from the ZAMG can also be mentioned as the most informative and action-oriented sources for the "Fire Brigade and Civil Protection" department. The flood information received from the hydrology department and the ZAMG during a flood event is processed professionally and in a structured manner by the staff of the fire brigade and civil protection department. The civil defence siren is another important component of the warning system and is the responsibility of the Fire and Civil Defence Department. However, this requires that the department, as well as external institutions, raise public awareness of the significance of a siren signal and the measures that must be taken afterwards.

E3 also explains that, on the Danube, the state warning centre generally distinguishes between two water levels. A distinction is made between "warning" and "alarm" measures, which are based on municipality-specific water levels and are specified in the special alarm plans depending on the threat. When a warning is issued by the state warning centre, all responsible and affected authorities are informed by telephone. When the alarm is sounded, operational, municipality-specific measures must then be taken. The measures to be taken should be laid down in special alarm plans. Measures that can be implemented directly by the state warning centre include coordinating scarce resources, such as the deployment of helicopters. E3 also states that two districts (Tulln and Vienna surroundings) in Lower Austria are excluded from this nomenclature and do not have a warning level. Over time, residential areas have been established in the floodplain of the Danube ( ). These are exclusively second homes, and the residents are aware of the danger. Since the warning level is already declared at a very low water level, experience has shown that a flood warning would be issued on average ten times a year in these two districts, which would cause unrest among citizens outside these districts (E3: lines 92–215).

Expert 4

E4 from Upper Austria explains that the most relevant information required for the flood forecast model is the precipitation data and forecasts provided by ZAMG. He also mentions the water level measuring stations, which provide key input data for flood forecasting. E4 also mentions that a hydrological model transitions into a hydrodynamic model on the Danube.

E4 states in the audio recording: "The hydrological model is relevant in the catchment area. This means that the precipitation that has fallen is the input and at this point I get a flow value. Then you add a certain transit time until the water actually reaches the Danube. From there, it is then *calculated* further using a hydrodynamic model" (E4: lines 67–71).

E4 also mentions that during a flood event, the amount of information received is not significantly higher than during dry weather. The information processing procedure is therefore always the same. The data from the individual measuring stations is transmitted to the hydrography department via modem, GPS or radio. In this department, the data is first stored on a server. The data is then "retrieved" from a data centre and processed in a data processing programme. Finally, this data is published on the web and fed into the forecasting system. In addition, the forecasting system is monitored by a control system in the background ( ). The "Views" system used by Deltares automatically corrects calculated values if they do not match the measured values. A preliminary alarm is triggered when a specified threshold value is reached on the Danube.

E4 explains: *"These are mostly water level values that are relevant. Because it has been said that flow values can only be reconstructed after a certain amount of time"* (E4: lines 87–89).

Fig. 37: Bubble sensor (source: www.ott.com)

A so-called pre-alarm is issued in Upper Austria by the Surface Water Department, which also includes the Hydrography Department. As soon as a value exceeds the pre-alarm threshold, a team consisting of experts from various emergency services and the state government is convened. With regard to the water level measuring stations, E4 mentions that different device systems are used to measure the water level in the Upper Austrian section of the Danube. A distinction is made between water level measuring stations with a pressure probe, as shown in Figure 37, and a bubble sensor, which is shown in Figure 38.

Fig. 38: Pressure probe (source: www.ott.com)

The state of Upper Austria also employs water level readers and technicians to optimise the data and recalibrate the water level measuring stations on a case-by-case basis. The water level measuring stations on the Danube are owned by VIADONAU, which provides hydrographic services for shipping during low and medium water levels. VIADONAU has decided not to be involved in flood forecasting and prediction and has left this task to the state.

As in Lower Austria, the state of Upper Austria has installed partially redundant systems in the water level stations so that data can be compared. This means that data transmission cannot fail (E4: line 52 - 183).

10.2.1 Summary

In principle, the experts surveyed can be divided into two groups. The first group, the hydrological departments of the respective state governments, are directly involved in flood forecasting systems. They work with different measurement data obtained from various sources. Their tasks therefore also include the evaluation and decoding of highly complex data in order to publish it in a simplified form that is understandable to the population. In the background of the data processing, numerous server programmes run, which store and process the data so that it can then be made available to the population in a transparent manner. The other experts are indirectly involved in the flood systems. They do not contribute to the preparation of flood forecasts. Nevertheless, they are responsible for ensuring that information is communicated and organisational processes run smoothly in the event of flooding. Various organisations are responsible for providing services in this regard: On the one hand, the population must be made aware of the risks through training courses, brochures and workshops provided by the civil protection association at . On the other hand, it is the task of a department specifically designated for this purpose, such as the "Fire Brigade and Civil Defence" department in Lower Austria, which in turn trains all those responsible, including the emergency services and authorities. Accordingly, the expertise of the second group is based on different sources of information than that of the first group. While the experts in the hydrology departments are also responsible for maintenance and calibration of the water level measuring stations during normal operations, the experts in the other two departments are constantly working on continuous improvements to special alarm plans at the district and municipal levels. During a flood event, these plans are crucial for smooth staff coordination and operational work directly at the scene, as they contain all the necessary measures that must be taken at the municipal level.

10.2.2 Interpretation

Cooperation between the hydrological services and the ZAMG must function smoothly, as otherwise it would not be possible to process the data efficiently and effectively for a flood forecast model. The high volume of input data into the flood forecasting system increases complexity. Since the ECMWF model produces forecasts with a range of up to ten days, the error rate is higher than in the hydrological model. Since the precipitation forecast data is incorporated into the flood forecasting model, incorrect data has a negative impact on the result. The error rate of the hydrological model is proven to be lower because the state employs its own water level and precipitation measurement observers. In the event of incomprehensible or missing data, this can be corrected by contacting the observer by telephone at . Hydrologists responsible for flood forecasting must be comprehensively trained and have the relevant experience and expertise. This also means that these experts must have a particular understanding of precipitation, water level and flow values. Comprehensive IT skills are mandatory.

10.3 Use of data

Expert 1

E1 states in the interview that flooding always takes on different dimensions due to the varying spread, water volume and flow velocity, and is therefore very difficult to predict using the same model. As a result, floods are classified according to a certain "annuality" based on statistical data.
E1 also mentions: "And that's where you talk about an HQ *100 [...]. You say that a flood of* this magnitude and intensity is expected in probability

calculations, which is again risk management, that *it happens every 100 [...] years"* (E1: lines 143 - 147)..

The statistics are constantly updated based on experience and changes in the terrain, as the dimensions of floods have changed significantly over time, for example due to weather phenomena. The statistical values are displayed graphically, as are the relative values, in the "NÖ Atlas". This programme offers a variety of display options, providing detailed information on which areas will be affected and at what levels. E1 also tries to explain how important it is to put these representations and values in writing in a disaster control plan and explains: *"[...] but we also know that 80% of disasters deprive us of electricity [...]"* (E1: lines 168 - 169). In its training courses, the civil protection association therefore repeatedly draws the attention of district and municipal representatives to alternatives in disaster control planning, such as always having plans printed out and ready, being able to call up water levels by telephone and familiarising themselves with radio (radio network) as a means of communication (E1: lines 139–173).

Expert 2

Our data is published as raw data on the Internet, on teletext and on a voice server, among other places, and is thus made available to the public transparently and free of charge. However, the only information available on the Internet is an Excel table with a time series for the last month or a graph with values for the last year. The water levels are five-minute values and the precipitation measurements are 15-minute values. Accordingly, a large amount of data is reported. Data for the more distant past is therefore only available in the state archives. However, if the Department of Hydrology and Geoinformation receives orders that exceed its daily workload, such as recording certain water levels for the last 20 years, it must charge for this at a prescribed rate. After another two years and many processes and measurements, the data is also published in a hydrographic yearbook. This yearbook is made available to the public as a download on the website of the Federal Ministry of Agriculture, Forestry, Environment and Water Management.

For the flood forecasting model, however, the raw data is processed by special databases and servers .

E2 explains: "Once a day, the system is checked to see whether the data is traceable, whether the forecasts are reasonable, or whether there are any serious deviations *that are not plausible"* (E2: lines 306 - 308).

If there are no deviations, automatic forecasts are adopted and published. In the event of flooding, however, forecasts are not published automatically; instead, each forecast is checked for plausibility.

A statistical evaluation of the data is carried out in a separate department of the state (E2: lines 308–381).

Expert 3

E3 mentions that the data obtained by the "Fire and Civil Protection" department from the "Hydrology and Geoinformation" department is used to protect people, animals and property. As the State Warning Centre makes up the majority of the "Fire and Civil Protection" department, the distribution of this data is its top priority. E3 therefore explains that the distribution network created by the State Warning Centre, consisting of all district authorities, municipalities and emergency services, is the user of this data (E3: lines 224–226).

Expert 4

E4 explains that, in addition to flood forecasts, the hydrographic department of the province of Upper Austria also produces flood reports and expert opinions. E4 also mentions that specific information tailored to individual citizens can be obtained from the hydrographic service by the citizens themselves. In the event of flooding, the hydrographic department of the province maintains close contact with the municipalities that have a mobile flood protection system. It is particularly important for these municipalities to receive accurate data, as they rely on the province's forecasts (E4: lines 184–196).

10.3.1 Summary

In principle, all services relating to flood forecasting are available to the public free of charge and at any time. In normal operation, the data is not subject to any special checks before publication and is therefore distributed automatically. In the event of a disaster, however, each individual data record is checked in order to avoid forecasting errors that could have serious consequences for the establishment of various protection systems. Training courses are held for citizens and emergency services personnel to help them understand the data prepared by the hydrological departments of the federal states. This ensures that the right actions can be taken in the event of a disaster. It is essential to draw up special alarm plans for each municipality, and to ensure that these are understood and applied. The civil protection association and the fire and civil protection department provide advice and numerous training programmes for this purpose. The state warning centre plays another important role in the hydrological department. It is responsible for ensuring that warnings and alerts reach every single municipality affected and the emergency services responsible for them.

10.3.2 Interpretation

In addition to warning and alerting, a distinction can also be made between the news service and a reporting system. Unlike the reporting system, the news service is carried out by the hydrological departments of the federal states, while the reporting system is outsourced and subsequently

Other departments in the state are responsible for this. Another key difference is the different areas of responsibility in specific situations. While the intelligence service carries out its work as normal, the reporting system only kicks in during floods. The intelligence service also supports the hydrology department by taking over the task of reporting water levels to other users. In Lower Austria, reporting is carried out by the fire brigade and civil protection department. This involves sending reports at certain intervals using standardised forms. The fire brigade and civil protection department then adds this data to the hydrological department's reports, which are updated every five to 15 minutes, and sends them to all specified users via a distribution list.

10.4 Decisions and financing

Expert 1

E1 describes: "Decision-making is purely a matter for the authorities. We train them" (E1: line 175). However, the authorities responsible are often unaware of this because the work and organisation during a flood event is mainly carried out by the local fire brigades. It is therefore the task of the civil protection association to raise awareness among the authorities and local representatives and make it clear to them that they also bear responsibility.

E1 adds: "This means that the mayor of a municipality is the head of operations. Even if the fire brigade takes over the operational part, the mayor *is ultimately responsible for the decisions* made *at* " (E1: lines 179–182).

The civil protection association is primarily funded by municipalities and the state. The association also organises events in which sponsors can participate. Since the tax reform, the association has been subject to taxation. The association is subject to the small business regulation. Donations are therefore subject to taxation above a certain amount (E1: lines 175–201).

Expert 2

E2 states that it is often a very difficult decision as to which forecasts to publish. With regard to the values and data available for flood forecasts, hydrologists must also estimate the margin of error. This makes decisions correspondingly difficult. Since much is handled by framework programmes during normal operations, numerous checks by hydrologists are required in the event of a flood to minimise the degree of misinterpretation.

The hydrological department is funded by the state and, under certain circumstances, also by companies that rely on forecasts (E2: lines 329–351).

Expert 3

E3 also mentions that, according to the Civil Protection Act, much of the responsibility lies with the district administrator. However, the "Fire Brigade and Civil Protection" department takes measures to coordinate and address resource shortages. Decisions are made at the state or district level in accordance with the principle of subsidiarity.

E3 explains: "Attempts are made to solve the problem where it arises *[...] and not three levels above"* (E3: lines 240–242).

In order to further improve the effectiveness of combating Danube flooding ( ), the "Fire Brigade and Civil Protection" department is committed to improving the "process understanding" of the Danube. This means that E3 considers it particularly necessary to inform a larger group of people about the behaviour and characteristics of the Danube when water levels are high. He would like to implement this in the form of training courses for representatives of the emergency services. The Fire and Civil Protection Department is financed from the state budget (E3: lines 229 - 259).

Expert 4

E4 explains: *"99% of our decisions are long-term because they have to be so well-founded that [...] no wrong decisions are made in the event of flooding"* (E4: lines 202–203). The hydrographic departments of the Upper Austrian provincial government are financed by the provincial budget (E4: lines 202–208).

10.4.1 Summary

Decisions in all the institutions mentioned are based entirely on the long term. Short-term decisions are mainly taken at municipal and district level. It should be noted, however, that in the event of a flood, computer programmes alone do not contribute to decision-making. Which data is ultimately made available is the responsibility of a small number of experts. Based on these data-supported interpretations and the flood forecasts derived from them, numerous measures are taken that are often difficult or even impossible to reverse. The flood-related services mentioned are largely financed by the state.

10.4.2 Interpretation

All the authorities involved make a significant contribution to combating flooding and minimising damage to property and personal injury. The responsibilities of the individual organisations are also clearly defined. The hydrological departments are at the beginning of the decision-making chain. They must therefore work very carefully and deliver high-quality work. Despite the support of state-of-the-art computer programmes, hydrologists must interpret this data. This decision-making process can have both positive and negative effects.

CHAPTER 11

Conclusion and outlook

Floods are the most frequent and damaging natural disasters in Austria. They can be exacerbated by climate, land use and hydraulic engineering (see BLÖSCHL et al. 2013: 9). However, heavy precipitation events are considered the highest risk factor for flooding (see HOFSTÄTTER 2011: 24).

In the course of this thesis, the research question was formulated and thus induced as to whether climate change contributes to the increased occurrence of flooding. In Chapter 6, the author of this thesis substantiates the hypothesis that climate change leads to an increased occurrence of extreme weather conditions, which have already triggered numerous flooding events in Austria in the form of precipitation.

In recent decades, there has been an increase in flooding in Austria, particularly in areas north of the main Alpine ridge (cf. BLÖSCHL et al. 2001: 29). Various documents show that these massive precipitation events were largely triggered by Vb-type storm tracks (cf. HOFSTÄTTER 2011: 26).. The influence of climate change, which is mainly caused by humans, cannot be ruled out. Anthropogenic influences on the climate system, including emissions of radiative greenhouse gases, are causing global warming. In Austria, this has already led to an increase in ground-level air temperature of approximately 1.2°C. This suggests

that climate change is causing a change in atmospheric water vapour, which in turn is leading to a change in potential precipitation. A specific weather classification system ( ) can therefore be used to illustrate the relationship between atmospheric water vapour and regional precipitation (cf. HOFSTÄTTER 2011: 67). A change in the frequency of certain weather conditions can therefore be cited as an indication of an increase in heavy

precipitation. In summary, there are two indications of a future change in the frequency of heavy precipitation. The first is an increase in atmospheric temperature and humidity, and the second is changes in atmospheric circulation (cf. HOFSTÄTTER 2011: 26).
Due to a possible increase in the risk of flooding, which is being driven by climate change, among other things, technical flood protection and mobile systems are to be mentioned as the core elements of all measures (cf. BLÖSCHL et al. 2013: 9). In addition to technical flood protection, preventive measures are also part of integrated flood prevention. Runoff forecasts are necessary in order to react in time in an emergency. The establishment of mobile flood protection in particular depends on timely information from hydrographic services, as these activities are very time-consuming (cf. BLÖSCHL et al. 2013: 11).

In Chapter 8, the author of this paper answers the second research question: "What factors are necessary for mobile flood protection to work?" The hypothesis that mobile flood protection only works in conjunction with flood forecasting is discussed in detail in Chapter 8.2.1. This chapter presents a detailed table showing the phased implementation of the mobile flood protection elements in the municipality of Weißenkirchen, which, due to the water level data it contains, can only function in cooperation with the flood forecasting service of the hydrographic department in Lower Austria. This is because, in order to begin setting up the mobile protection system , the municipalities affected are dependent on water level and flow forecasts from the hydrological services (cf. KREMS DISTRICT ADMINISTRATION 2014: 8).

Of particular importance is the classification of the hydrological services' reporting system, which is divided into three levels (cf. KREMS DISTRICT ADMINISTRATION 2014: 8ff.):

- Early warning stage: "Kienstock" level is approx. 5.50 m.
- Flood warning level "Kienstock" is reached at 6.15 m and
- flood alarm is triggered at a "Kienstock" level of 7.50 m.

The individual construction phases of the mobile flood protection system are closely linked to this classification. This only works if the system is set up correctly, with a sufficient number of construction workers and coordination of these workers. Special exercises and training courses are held annually with the Austrian Armed Forces for this purpose. The army is available to provide support in the event of a flood (DENK: 2016). The author of this thesis took part in an operational exercise to learn about the details of setting up the mobile system in Weißenkirchen. The exercise lasted a whole day and provided insight into the many technical challenges that must be overcome for the mobile flood protection system to function.

Due to the fact that mobile flood protection systems function exclusively on the basis of hydrological flood forecasts and a corresponding advance warning time, the third and final research question, "What functions do flood forecasting systems and early warning systems fulfil?" is answered (cf. BML-FUW 2002: 33). The author substantiates the corresponding hypothesis that computer forecasts and weather calculation models can be used to increase the advance warning time and thus reduce damage, on the one hand with the help of qualitative guided interviews and, on the other hand, through the theory-based Chapter 7. For an adequate verification of the aforementioned hypothesis, the interview guide used was categorised. This should facilitate the evaluation of the interviews and thus improve the results.

In summary, it was mentioned several times in the thesis that flood forecasting is the provision of timely and effective hydrological and meteorological information (cf. BLÖSCHL et al. 2013: 9). Flood forecasting and the preparation and follow-up of flood events are the responsibility of different institutions. In order to ensure the functionality of flood-related institutions, powerful computer programmes and trained and qualified personnel are required (cf. BLÖSCHL 2008: 41). In the hydrographic departments of the state governments, the responsible persons interviewed by the author are graduates of the University of Vienna and the University of Natural Resources and Life Sciences. During their studies, they specialised in hydrology in order to be able to fulfil the tasks of their current jobs (E2: lines 3–4; E3: lines 75–78; E4: lines 32–33). At the Lower Austrian Civil Protection Association, on the other hand, great importance is attached to experience. The current managing director of this association is able to apply the knowledge he acquired during his many years of operational service in the Austrian Armed Forces to his current field of work (E1: lines 28–57).

However, the municipalities concerned, the district authorities and the citizens themselves also bear a great deal of responsibility. They are directly affected and are therefore ultimately the ones who must take measures on the ground to protect against flooding. The most important aspect of modern, integrated flood management is therefore the effective cooperation between all these institutions and the citizens involved (cf. BLÖSCHL et al. 2013: 9).

In principle, the experts interviewed can be divided into two groups. One group is directly involved in flood forecasting systems. They are responsible for evaluating, interpreting and providing various highly complex measurement data. The other group is indirectly involved in flood forecasting systems. Their tasks include information transfer and organisational processes before, during and after a flood event. Despite their different areas of responsibility, these two groups must work together smoothly to minimise the potential damage caused by flooding (E1: lines

62–85; E2: lines 62–218; E3: lines 92–166; E4: lines 56–78).

Although the hydrographic departments of the federal states provide their services to the population free of charge, these are often not taken up due to lack of interest or a lack of understanding. This is because the importance of systematic and comprehensive flood protection is usually only recognised by citizens after a flood has occurred and in view of the consequential damage. In order to raise awareness of this issue among the population, the state offers various training courses and exercises, most of which are also funded by the Federal Ministry of the Interior ( ). These training programmes are relevant for emergency services and authorities on the one hand, and for citizens on the other. The implementation of these projects requires cooperation between the state government and the civil protection association (E1: lines 12–25).

In addition, the data processed in the flood forecasting model should encourage the creation of special municipal emergency plans. These plans are closely linked to flood forecasting systems and mobile flood protection systems. They include all areas at risk, all warning and alarm levels and the measures to be taken in each case.

These plans also make a significant contribution to minimising damage during flood events by coordinating emergency services and assigning tasks. Due to their complexity and the amount of work involved, only a few municipalities have drawn up special alarm plans to date. However, these municipalities benefited from their alarm plans and thorough preparation during past disasters (E1: lines 71–85).

The federal states support the creation of special alarm plans in Austrian municipalities with potential hazards through special funding programmes and training courses. The aim is to make municipalities and citizens more aware of the importance of property protection, personal responsibility and risk awareness (DENK: 2016).

Nevertheless, it is ultimately those directly involved in the flood event who must respond to the interpretations and results of the hydrographic services' assessment of the situation and make operational decisions. As this complex process can lead to incorrect assessments, municipalities, district authorities and citizens must also be able to quickly find alternative solutions (DENK: 2016).

The services and support programmes are funded exclusively from the state budget and must be evaluated on an ongoing basis through research results (E1: line 209; E2: line 404; E3: line 263; E4: line 2013).

Ultimately, several factors are key to preventing flood damage. Firstly, preventive measures for those affected in the form of exercises and training courses should raise awareness and understanding of flood events among the population. Functional forecasting systems should also increase the preparation time for protective measures. This leads to the conclusion that event-related information must also be communicated in a timely manner and in full. Mobile flood protection systems are also necessary because, on the one hand, there is often no space for technical flood protection (e.g. dams, retention basins, etc.) and, on the other hand, mobile flood protection must be installed as an additional protective measure due to higher water levels and flow rates caused by climate change. In addition, special alarm plans are becoming increasingly important in municipalities and district authorities.

CHAPTER 12

Bibliography

OFFICE OF THE LOWER AUSTRIAN PROVINCIAL GOVERNMENT (publisher) (2016): Flood Alarm Plan. Draft. St. Pölten.

OFFICE OF THE UPPER AUSTRIAN PROVINCIAL GOVERNMENT (ed.) (2013): Flood disaster 2013: review - improve - protect: How flood forecasting works in Upper Austria. Press conference held on 6 August 2013 in Linz.

KREMS DISTRICT ADMINISTRATION (BH KREMS) (2014): Special Alarm Plan. Danube Flood. -Krems.

BLÖSCHL G. (2008): Flood warning – on the value of local information. In: Intl. J. River Basin Management 6 (1), 41-50.

BLÖSCHL G. (2015): Changes in high water levels in the context of global change. -In: German Research Centre for Geosciences. -Potsdam.

BLÖSCHL G., NESTER Th., KOMMA J., PARAJKA J. and PERDIGAO R. A. P. (2013): The June 2013 floods – analysis and consequences for flood risk management. -In: Austrian Engineering and Architecture Journal 158 (1-12), 1-11.

BREMICKER M., HOMAGK P. and LUDWIG K. (2006): Early flood warning and flood forecasting in Baden-Württemberg. -In: Wasserwirschaft 7/8, 46-50.

FEDERAL MINISTRY OF AGRICULTURE, FORESTRY, ENVIRONMENT AND WATER MANAGEMENT (BMLFUW) (ed.) (2001): Guidelines. Procedures for estimating flood parameters. -Vienna

FEDERAL MINISTRY OF AGRICULTURE, FORESTRY, ENVIRONMENT AND WATER MANAGEMENT (BMLFUW) (ed.) (2002): Protection against natural hazards in Austria. -Vienna.

FEDERAL MINISTRY OF AGRICULTURE, FORESTRY, ENVIRONMENT AND

WATER MANAGEMENT (BMLFUW) (ed.) (2006): Freibord - Überströmstrecken (Free board - overflow distances). Guidelines for determining the required free board based on project-specific conditions, including criteria for the arrangement of overflow distances. -Vienna

FEDERAL MINISTRY FOR TRANSPORT, INNOVATION AND TECHNOLOGY (BMVIT) (ed.) (2004): Analysis of the flood events of August 2002 - FloodRisk. - Vienna.

BUREK P., THIELEN-DEL POZO J., THIEMIG V. and SALAMON P. (2011): The European Flood Early Warning System (EFAs). -In: Hydrology and Water Management 4, 193-199.

DENK M. (2016): verbal information dated 2 March 2016. -Weißenkirchen.

DONAU CONSULT (2006): Overall river engineering project for the Danube east of Vienna. -Vienna.

DYCK s. and PESCHKE G. (1983): Fundamentals of Hydrology. -Berlin.

EU DIRECTIVE (2007): Directive 2007/60/EC on the assessment and management of flood risks. E.P.u. Council.

FLICK U. (2002). Qualitative social research. An introduction. Reinbek near Hamburg.

FLICK U. (2011): Qualitative Social Research. An Introduction. Hamburg.

FORMAYER H. and KROMP-KOLB H. (2006): Floods & Climate Change. Effects of climate change on flood events in Austria. BOKU-Met. - Vienna.

FREYTAG-BERNDT and ARTARIA KG (2008): Austrian School Atlas. - Vienna.

GLÄSER J. and LAUDEL G. (2010). Expert Interviews and Qualitative Content Analysis. Wiesbaden.

GODINA R. (2006): Flood forecasting and warning – experiences from from the perspective of the HZB. In: GUTKNECHT D. (ed.): Flood forecasting. Experiences, developments & reality. Vienna. (=Vienna

Reports. Water, Wastewater, Water Bodies 199), 133-141.

GRILL R. (1950): Geological history of the Danube region in Austria. -In: Universum 14(15/16), 458-464.

HYDROGRAPHIC SERVICE OF LOWER AUSTRIA (HD NÖ) (ed.) (2016): Flood forecast for the Danube. -Lecture given on 18 February 2016 in St. Pölten.

HOFSTÄTTER M., JACOBEIT J., HOMANN M., LEXER A., CHIMANI B., PH LIPP A., BECK C. and GANEKIND M. (2015): WETRAX. Weather Patterns, Cyclone Tracks and related precipitation Extremes. Effects of climate change on large-scale heavy precipitation in Central Europe: Analysis of changes in tracks and general weather conditions. (Geographica Augustana 19), 8-70.

HUPFER P. and KUTTLER W. (eds.) (2006): Weather and Climate. An Introduction to Meteorology and Climatology. -Wiesbaden.

JUNGWIRTH M., HAIDVOGEL G., HOHENSINNER S., WAIDBACHER H. and ZAUNER G. (2014): Austria's Danube. Landscape - Fish - History. - Vienna.

KEILER M. and THALLER D. (2009): The Danube. Lifeline of a metropolis. - In: EMBLETON-HAMANN C., KEILER M. and TEUFL I. (eds.): Vienna - Environmental City Guide. Insights into the nature of a big city. - Vienna et al. 43-49.

KOMMA J., RESZLER Ch., BLÖSCHL G. and HAIDEN T. (2006): Ensemble forecasts of flood discharges. -In: GU KNECHT D. (ed.): Flood forecasting. Experiences, Developments & Reality. -Vienna. (=Vienna Communications. Water, Wastewater, Water Bodies 199), 279-293.

NIEHOFF D. (2001): Modelling the influence of land use on flood formation in the mesoscale.
-University of Potsdam, Potsdam.

ÖNORM B 2400 (Austrian Standards Institute) (ed.) (2004): Hydrology - Hydrographic terms and symbols - Supplementary provisions to

ÖNORM EN ISO 772. -Vienna.
PATT H. (ed.) (2001): Flood Handbook. Effects and Protection. Berlin, among others.
REINDERS H. (2005): Conducting Qualitative Interviews with Young People. A Guide. -Munich.
SCIETEC (2006): Flood Forecasting for the Danube. Documentation.
SEIBERT P., FRANK A. and FORMAYER H. (2006): Synoptic and regional patterns of heavy precipitation in Austria. (Theor. Appl. Climatol. 000), 1-15.
WEIßENKIRCHEN (2016): Material overview – Flood protection. -Weißenkirchen.
WEIßENKIRCHEN ALARM PLAN (2016): Warning and alarm plan. Danube flooding. -Weißenkirchen.

CHAPTER 13

Internet

FEDERAL MINISTRY FOR TRANSPORT, INNOVATION AND TECHNOLOGY (BMVIT): https://www.bmvit.gv.at/verkehr/schifffahrt/hochwasserschutz/index.html, accessed on 10 March 2016.

LOWER AUSTRIA ATLAS (NÖ ATLAS): http://atlas.noe.gv.at, accessed on 6 April 2016.

Appendix

17.1 Transcript of expert 1

Interviewee:	Thomas Hauser
Date	15
Time	10:00 a.m.
Location	Lower Austrian Civil Protection Association, Langenlebarnerstraße 106, 3430 Tulln
Participants	I: Julia Hasenhütl E1: Expert 1

I: How is the Lower Austrian Civil Protection Association organised?

E1: The Lower Austrian Civil Protection Association is an association with its own statutes. It is

we also receive orders from the state government, of course, and un-

our fundamental task is to inform the population in the area of crisis and disaster management and, above all, to encourage and motivate people to protect themselves, so that every citizen can make their own contribution and, in the truest sense of the word, does not have to wait for the emergency services to arrive, but takes early action to minimise damage.

I: What is your main personal role in this association?

E1: I am the future regional director of the Lower Austrian Civil Protection Association.

I: To what extent do you cooperate with others in your daily work?

E1: Very intensively with the state, of course. First of all, the

The state also finances this association with a corresponding budget. In addition, there are also funds from the municipalities, which make a kind of membership contribution. This finances our organisation, which consists of eleven full-time employees and around 1,200 volunteers in Lower Austria. We are the largest association in Austria, with associations in all federal states. Cooperation with the state government is fully guaranteed. Firstly, due to our proximity to the fire brigade and civil defence department, which is located directly on the floor below us. This means that we regularly have joint projects. There are also official notices that require us to carry out certain activities, such as training in crisis and disaster management at the municipal level, i.e. for municipal representatives and mayors, which was assigned to us by official notice. This means that the state government has tasked us with providing this training and support for the training.

I: What training did you have to undergo to perform your duties?

E1: There are various aspects to my job. First of all, I come from military service. I served in the Austrian Armed Forces for around 16 years. I am a trained non-commissioned officer. I work in the civil protection association, where we also have a civil protection school and have completed all the courses, including advanced and compulsory courses. And what is a very important aspect for us is experience. The work at the civil defence association is very difficult to categorise into a collective, as it is very, very diverse. It starts with giving presentations, moderating, dealing with children and young people, and of course also dealing with older people, the pensioners in our area. The topics range from radiation protection and flooding to heavy rainfall and a wide variety of other areas. It's just very, very diverse. And I have to say, to be completely honest, that I certainly did most of my training in the military, including leadership procedures and all other areas. What I can add as personal experience when talking about disasters is that I can definitely draw on my own experiences. I live in Hadersdorf am Kamp, and the floods in 2002 were an event where I was not only there as a helper, i.e. as a soldier on assistance duty, but was also definitely personally affected. My living room was also flooded. And that is, of course, experience that can be very well incorporated into our information work. I came to the civil defence association via a detour through voluntary work. That means I spent around ten years volunteering for the civil defence association, as a speaker, as a district as managing district director. And in 2007, I finally joined the Civil Defence Association as a full-time employee. I started as a regional manager, then became deputy state secretary, and since the new statutes came into force, this position of deputy state secretary no longer exists; instead, a state managing director has been appointed. I also said "future" at first. This is because we will have our first state executive committee meeting on 25 February, where the appointment will be made official. That is why I said "future" and am currently acting as interim state managing director.

I: What resources and tools are available to you to fulfil your duties?

E1: Specifically for the early warning systems for flooding?

I: Yes, exactly!

E1: It should be noted that there are two parts to this. The first, of course, the processing of the data. This is done by the Department of Hydrology and Geoinformation of the Province of Lower Austria. In a very exemplary manner. I believe that the transparency with which this data is made available is unique in Austria. The next point is, of course, motivating the population to use this data. It is also our job to motivate people and, of course, to show them where this data is available and where they can find more information. And in the event of an incident, it is also important to be warned in good time, which brings us to the siren signals and civil defence signals that are used here. And what lies between these two areas and is of course also an issue for us is disaster control planning at municipal level. This means that every municipality must address this issue. Am I affected by this or not?

also an issue for us is disaster control planning at the municipal level. This means that every municipality must address this issue. Am I affected by it or not? What steps do I need to take, who do I need to inform and when? In Lower Austria, this also extends to special alarm plans, and the disaster control plans are of course also based on the plans developed by the federal government. We are also preparing a wide range of guides to keep the population informed about this issue. As I mentioned at the beginning, the problem is often that citizens tend to ignore this type of danger and believe that there must be someone else who will take care of the problem. There are areas and authorities that deal with this, but that does not relieve us of our responsibility to protect ourselves.

Because in the event of a disaster, it is a concentrated load that arrives relatively quickly over a large area, and it is necessary for everyone to take the right steps. That is why I should familiarise myself with this in advance.

I: How do you obtain current water levels and water level forecasts?

E1: We obtain these from the Hydrology and Geoinformation Department based in St. Pölten.

I: What measuring points do you use to predict water levels?

E1: The state operates water level gauges and precipitation gauges to measure precipitation.

E1: Where are the measuring stations located on the Danube?

I: Where are the measuring stations located on the Danube?

E1: They are shown in detail on the "NÖ Atlas". The individual points shown

on the "NÖ Atlas" are measuring stations. Of course, there are also manual measuring stations. These are the slats that you often see. These are additional measuring stations. However, electronic transmission is displayed via these measuring points in the "NÖ Atlas". If you click on such a point in the "NÖ Atlas", you will see an actual value. This is updated every 30 minutes.
, you will see an actual value. This is updated every 30 minutes.

I: How many of these are currently in use?

E1: All of them.

I: Are there different water level stages?

E1: That depends on the levels. If I take my home parish as an example my home town. The disaster control plan naturally also shows a special alarm plan for flooding. If I draw up a special alarm plan, regardless of the event or scenario, I need a marker. I need a value. If you take the municipality of Hadersdorf am Kamp, we have various barrages in the upper reaches of the Kamp. There is Thurnberg, there is Ottenstein, there are various barrages that have been built into the Kamp over the years and have protected the municipalities from flooding for many years. Hadersdorf has taken the discharge volume from Thurnberg as the value for its special alarm plan. This means that when 30 m^3 of water passes through Thurnberg, the telephones start ringing at least once and those affected in the local operations centre are informed. If this continues to 50 m^3, the first measures are initiated. That's when the operations centre starts planning. For example, where do I need to close a section of road, or where do I need to inform people to clear their cellars, and so on. And so it escalates until the entire population is alerted and the community is evacuated. That is how a special alarm plan is structured, and these measurements are, of course, an absolutely essential factor.

I: Are there any countermeasures when an early warning is triggered? (Closing sluices, sandbags, alerting emergency services, etc.)

E1: That depends on how far the construction work has progressed . Let me come back to my home town. Hadersdorf is in a situation where you have the large Kamp river flowing past on one side and the Mühlkamp flowing through the middle of the village. Above, in the cadastral municipality of Kammern, the Kamp is dammed and the Mühlkamp is filled via this weir. As the name suggests, this is a body of water that was designed for the mills and to operate them. In 2002, it was not possible to seal it off again. Now there is this bulkhead gate on the federal road, where you can now seal off this gate, and with that I can prevent a large part of the municipality from being flooded. If these emergency measures are in place, there are also plans for them. When does what have to happen at what point in time? It's comparable to the mobile flood defences along the Danube. Whether it's Stein, Weißenkirchen, Dürnstein, Spitz or even the municipality of Rossatz on the other side of the Danube. They now also have a mobile flood defence system. And it is clear that when the water level is reached, for example on the Danube, construction must begin. At stage one, for example, the uprights and the first elements are installed. And depending on how things progress, this is then increased in stages. At this water level, that means stage two, until full construction is complete. Because flood protection planning is individual, each municipality has its own embankment height, each municipality has a different area layout, low points, higher points, does it go up at an angle, does it go more flatly to the rear, or does it slope downwards, so the flood areas are naturally also different. The risk of the flood coming relatively quickly is lower in one municipality, but in another municipality it is then very extreme.

171 I: What does the statistical analysis and regionalisation of the data look like?

173 E1: An HQ is a statistical value. This HQ also refers to a

174 Frequency. For example, a 100-year frequency. This is what it's about. 175 It is very difficult to measure a flood itself in a diagram because, due to its spread, the volume of water and the flow velocity, it always takes on different dimensions. 178 And this is represented in a statistical value. And that is where we talk about HQ 100, HQ 30, HQ 1 or various other HQs. You say that a flood of this magnitude, of this intensity, is expected in probability calculations, which is again risk management, to happen every 100, 30 or 1 year. The statistics are, of course, constantly being revised, especially for the Hydrology and Geoinformation Department. Because there have already been situations on the Danube where an HQ 1 occurred three times in six months. That means that would no longer be

accurate. That's why it's being revised and updated. This is a continuous process. But it is a way of specifying dimensions and illustrating effects. The "NÖ Atlas" therefore offers a variety of representations. On the one hand, I have a measured value that tells me how many metres^3 water is flowing in the customer's property through the water level, and on the other side I have a map overview that shows me, at an HQ 30, how far this has spread and what area is affected. This is often in the centimetre range. In this case, it must be said that this has to be represented by statistics and not 196 by a clear value. Clear values are then provided directly to the affected municipality by the hydrology and geoinformation department at the time of an event. There, it really matters, above all,

199 because it is crucial to know when to deploy mobile flood protection, when it will be set up, and how much lead time is available. There are also plans for this. The affected municipalities also need values and forecasts. The forecasts for the Danube, for example, go up to 48 hours. This is also essential assistance for protection and damage minimisation.

205 I: So your work is essentially for the purpose of disaster control planning?

207 E1: Yes, exactly. And it is also particularly important to have this in paper form in

208 of the municipality. Everything is wonderful as long as everything can be found on 209 the computer or on the Internet. However, we also know that 210 80% of disasters deprive us of electricity because 211 there is a power failure. This means that we have to fall back on alternatives in our disaster control planning that we also have at our disposal. That is why these plans are also printed out, why water levels are sometimes retrieved by telephone or transmitted via a radio information system, so that we can react correctly and quickly.

217 I: What are the implications of the decisions you may have to make?

219 E1: Decision-making is purely a matter for the authorities. We train them.

220 This is divided into the NÖKHG, the Lower Austrian Disaster Relief Act, which also clearly states that the authorities at their respective levels have their areas of responsibility in the event of a disaster. It is also often said that the fire brigades are responsible for this. However, the responsibilities clearly lie with the authorities. This means that the mayor of a municipality is the head of operations,

even if the fire brigade takes over the operational part, the mayor must make these decisions and ultimately be accountable for them. And that is also a major point where we focus our training in order to prepare for such situations.

227 and must ultimately be able to take responsibility for them. And 228 that is also a major point where we focus our training 229 in order to convey the in clear language and to make this clear to the mayors. Because the rude awakening usually comes 231 only afterwards.

232 I: What measures do you see to improve the effectiveness and efficiency of your services?

234 E1: Offering more courses to reach more citizens and thereby raise awareness among the population. We offer a wide variety of training courses. If we look purely at high-rise building or disaster control management, we have the information I mentioned, and there are two training modules. Module one deals with the contents of a disaster control plan. How should this be structured? What should this plan contain? What data is really necessary? This module training is mandatory for municipalities. Disaster control is state legislation, and the NÖKHG states in paragraph 14 that every authority must have a disaster control plan. Similarly, paragraph 10 states that the mayor is appointed as the local head of operations. So there is already a clear legal basis for this.

247 I: How do you finance your services?

248 E1: Primarily through local authorities and the state. We also organise events where we have sponsors. Unfortunately, however, due to the tax reform that affects us as an association, namely the small

249 events where we also have sponsors. Unfortunately, however, due to the tax reform that we as an association are subject to, namely the small business regulation, the sums that come in as donations are kept very low, because otherwise it would be a huge tax disadvantage for us. However, we do have regular partners, such as the "Safety Tour" we organise. For 16 years, we have been running safety tours, which are children's safety Olympics, where we teach children in the fourth year of primary school about road safety in a playful way.

255 Safety Tours for 16 years, which are children's safety Olympics where 256 we confront fourth-grade primary school children with this safety topic in a playful

way. For example, it's about increasing road safety on a bicycle course. 258 The ÖBB has been a partner of these Safety Tours for several years. There are special games, such as how to behave at train stations, at bus stops, where can I avoid dangers . That's where our information work actually begins. In kindergarten, we tend to focus more on the parents, whom we take under our wing. But in primary schools, we definitely focus on the children, and this continues with school presentations, evening lectures in the communities, panel discussions, various safety days, presentations and events of all kinds. Last year, in 2015, we held around 1,184 events with around 96,000 visitors.

17.2 Transcript Expert 2

Interviewee:	Hauer Franz
Date	18
Time	9:00 a.m.
Location	Lower Austrian Provincial Government, Landhausplatz 1, 3109 St. Pölten
Participants:	I: Julia Hasenhütl E2: Expert 2

I: What training did you have to complete in order to perform your duties?

E2: I am a geologist specialising in hydrogeology. That is why I came here as a hydrogeologist, as a groundwater expert.

I: Did you study in Vienna?

E2: Yes, with Dr Häusler, he's still there.

I: What is your professional background?

E2: I previously worked for three years as a civil engineer. a civil engineer. He then closed his office. Then I was self-employed for threeand then I found found my job here. About my training: I'm actually a geologist , but I've also done a lot of IT work computers, and this job involves a lot of IT. Let's put it this way, the hydrology you need for this is something you have to learn on the job , but that's the case everywhere. You can't get anywhere with just what you've learned. I've been working here since 2000.

15 I: Could you describe how the hydrology department is organised?

17 E2: Basically, I can tell you that we are the Hydrology and Geoinformation Department, and the hydrology section is divided into the Surface Water, Precipitation and Groundwater divisions. And then there is the Water Level Information Service, which I head.

22 I: What exactly is your area of responsibility?

23 E2: My job is to make flood forecasts. And

24 my specific area of responsibility as head of this department is to ensure that everything runs smoothly from an organisational and technical point of view. I am also an expert in hydrogeology.

27 So my job is mainly the organisational and technical management and maintenance of this operation. This ensures that in the event of flooding, the people who are actually on flood duty have the appropriate infrastructure and are able to use it.

31 This means that training is also part of my job.

32 I: To what extent do you cooperate with others in your daily work?

34 E2: You can distinguish between technical cooperation, i.e.

35 where we get our data, and informal cooperation, i.e.

36 what we do with it and how we publish it. Technically, we cooperate with

37 neighbouring countries, i.e. the Czech Republic, Slovakia, Upper Austria, Salzburg,

38 Bavaria and, to some extent, Styria. And then informally, we make the flood forecasts and then publish the whole thing. We don't just make flood forecasts, we publish the whole thing, i.e. the data, on the internet. We have online data.

42 In this respect, we communicate with everyone who has internet access, and

43 warnings go directly to the state warning centre. This means that we do not issue any personal warnings, nor do we issue warnings to municipalities, but everything goes to the state warning centre, which then distributes the warnings in accordance with the existing alarm plans.

48 then distribute the warnings according to the alarm plans in place.

49 This is then passed on to the municipalities via the district level. These are the official warning channels as provided for by the authorities. However, anyone can call us at any time and ask whether the information on the Internet is correct. The values can be found at www.wasserstand-niederoesterreich.at. There you will find all the measuring points we have, with the

52 measurements and forecasts, either in tables or as graphs or on a map. The measuring points that you find on the internet all work electronically . However, there are still a lot of manual measuring points, but these are not listed on the internet, of course. So, the hydrological service is originally about collecting and storing data.

57 So we are the hydrographic service of Lower Austria and we do this in indirect federal administration for the hydrographic central office in the ministry. They have more or less outsourced the activities, and that was the original task, which was to simply collect the data from the water level gauges and precipitation gauges, correct it if necessary, and then pass it on to where it is collected centrally. Later, they were equipped with electronic devices instead of writing instruments, and then remote data transmission was added. And then the question arose of what to do with the data that came in automatically, and it was decided to create forecast models and so on. That's how it developed, and it now exists in all federal states as a result of this development. However, we also cooperate with EVN due to the military facilities.

71 I: What resources and tools do you have at your disposal to fulfil your current tasks?

73 E2: We receive weather data from the ZAMG (Central Institute for Meteorology and

74 Geodynamics). We need to briefly mention the data flow we have in more detail. We send our measurement data from the precipitation measuring stations to the ZAMG. The ZAMG also has its own measuring stations. So we send them the data, they have their own data, they purchase the weather data, i.e. the weather radar data, from Austro-Control. Austro-Control has a few radar stations in Austria, primarily for aircraft and secondarily for weather. They have a lot of data and basically sell the weather radar data, only the radar data, to ZAMG. However, ZAMG also receives data from Europe-wide and global precipitation models, and with all this data, they create a precipitation distribution map optimised for Austria. Among other things, the ZAMG uses the ECMWF model, which is calculated in London. And what does the ZAMAG do? Because we only have point data where the measuring stations are located. The radar data is then distributed over the entire area. Radar data produces reflections from something. This means that if there is a cloud or fog somewhere, something reflects, there is a shadow on , and this creates an image. The ZAMG receives this and, based on the precipitation measuring points and various analyses and statistical evaluations, says, okay, this radar shadow with this signature means it's raining there, because I can check that with this measuring point. So

the measuring stations are, so to speak, the reference values for the weather radar. Only the weather radar covers a large area and the measuring stations are selective. This is very important in thunderstorms. A thunderstorm that appears very locally is not even noticed at measuring stations, or only very slightly. When we then analyse our measuring stations, we notice, aha, there

100 It only rained lightly, or not at all, and in reality, 101 people say there was a huge thunderstorm there. You can see that on the weather radar. That's the difference. Because I could actually take my measuring points and say, let's do an interpolation between the measuring points, and I would also get a result for the whole area, but it's not quite the same.

106 And then the ZAMG makes a forecast, a precipitation and temperature forecast that is of interest to us. And we then receive these precipitation forecasts from them. We receive them every 15 minutes or hourly, depending on what we need them for. And these precipitation forecasts are then the essential input data for the forecast models, for the flood forecast models. A precipitation forecast is structured like this. First, I have a so-called analysis. This is where the precipitation data comes in, i.e. the measuring stations and the weather radar data, and then there are endless evaluations that take years. This means that the weather radar is not weighted equally everywhere, but at certain locations, you know that the weather radar generally provides poorer coverage, which means that you have to take certain factors into account. So it's not a simple matter.

119 You look at the last hour and the current hour and how the individual cells, the precipitation cells, have moved. With a thunderstorm, this is easy to see, because I have a thunderstorm tower that can either move or stay in one place. Thunderstorm fronts tend to move quite a lot. So how have they moved in the last hour? And 124 that is simply moved forward using vectors. That is 125 completely taken over in the next two hours and in the four hours 126 after that, the whole thing is already overlaid by two weather models. One

127 one is the ECMWF, which covers Europe, and the second is the ALADIN DE 128 VIENNA. This is a French weather model, a small-scale model, and 129 the ALARO is a further development of the ZAMG and is specifically 130 geared towards the Eastern Alps. This means that it is the small-scale weather model and the ECMWF is the large-scale weather model. The large-scale model provides the input data for the small-scale model, but some of the data is also taken from the large-scale model. This is because there are

situations where rainfall is involved, and in these cases the large-scale forecasts are simply better than the small-scale ones, which is why we don't want to completely discard them. After 72 hours, the ALARO is also out and then only the ECMWF remains. The ECMWF then goes up to 10 days into the future.

138 And the timing is such that both ECMWF and ALARO are calculated twice a day, but offset by 12 hours. This means that a model is recalculated every six hours, which is why there are such large jumps in the precipitation forecasts. They are not continuous, but you know, okay, now it's ALARO or ECMWF. New forecasts come in from them and now it may be that the forecasts are somehow different. That part is always new, we get it every 15 minutes. INCA is all together.

146 That means I get a forecast, one that has been produced.

147 So there is the deterministic forecast or the main run, and then there is the ECMWF, which is simply shifted. That means that there are uncertainties in the model. It is not 100% certain that it will happen. It could also be different. Now you shift it randomly and that's how you get ensemble values. And that looks like this. I have some precipitation forecasts and that is the so-called main run, or the best forecast that was believed.

154 The actual observation may be somewhere else, because it's

155 all a forecast.

156 The precipitation is the decisive input data because the models all calculate in the same way in principle. It rains, or it will rain, then they have stored the soil moisture and other parameters. Snow water value and the forecast comes that it will start to thaw. The snow begins to melt and at the same time it rains. It rains on the thawing snow, causing runoff. Our precipitation gauges are automatic precipitation measuring devices.

162 Rainfall gauges are automatic precipitation measuring devices.

163 . They also have a temperature sensor in a standardised windshield. A so-called ombrometer. The precipitation is measured manually once a day. It is heated and the snow is therefore also measured as precipitation. There is also a scale inside that calculates the weight of the precipitation.

168 That's one thing. The other thing is that, as I mentioned at the beginning, we cooperate with other hydrographic services. For example, Upper Austria is worth mentioning because they are located upstream. But we also cooperate with the Czech hydrographic service, which is the CHMU (Czech Hydrometeorological Service), and with the HND (Flood Information Service) in Bavaria. We receive forecasts directly from

the latter. From CHNU on the March, we also receive forecasts for the Thaya in Austria, which has its own flood forecasting model.

174 the latter, we receive forecasts directly. From CHNU on the March, we also receive information for the Thaya in Austria, as they have their own flood forecasting model for the Thaya/March area. And there we get, especially for the Hohenau gauge, which is at the Zwickl Czech Republic-Austria-Slovakia border, where the Thaya flows into the March, we get a forecast and these forecasts are in turn input data for the March. And the same is available from Bavaria for the Danube.

180 data for the March. And the same applies to Bavaria for the Danube. The

181 Bavaria makes a forecast for the Passau water level, which is after the confluence of the Inn and Danube, and passes this on to us, and that is then the input data for our forecast. We then calculate down the Danube, i.e. through Upper Austria. Again with data from Upper Austria, whereby Upper Austria and we have the same model for the Danube. This is a model and can therefore also be calculated jointly. We calculate separately, but it is one model.

189 There are several forecast models in Lower Austria. We have the Danube forecast.

190 Danube forecast. The Danube model consists of two parts. We have a hydrodynamic model with 1144 cross-sections, with power plants and floodplains, and the tributaries are modelled by the precipitation-runoff model. Precipitation-runoff model: rain falls, the status is stored, and then the forecast comes with rain falling and so on. In the 1D model, only what happens to the water based on the cross-sections is calculated. The cross-section says that with such a flow, the water level rises by this much and the flow velocity changes in this way, and then an inflow is added at the next node. What happens then? Then we have something with EVN on the Kamp, although we don't publish these forecasts on an ongoing basis, only in the event of high water, because EVN doesn't want its operational management to be published. This is electricity data, and electricity data is expensive and of interest to the competition. Then we have something on the Traisen and the Ybbs. We have information from our colleagues in the Czech Republic about the Thaya. They send us the forecasts. March is actually a tributary, and we also get that from the Czechs, and the rest is modelled by us. There are 49 stations in total. Then there are the so-called secondary catchment areas, which

which are just small individual gauges that are not connected and are relatively easy to model in order to indicate an order of magnitude. In my opinion, they are also very inaccurate. The small catchment areas are not indicated by a hydrograph, only the order of magnitude. So, for example, the forecast or water level will lie between MQ and HQ1. Because this is usually so inaccurate that no one wants to publish anything more precise. At the moment, it is already relatively complete, and there is also a model for the Erlauf. This is already in the test phase, but it has not yet been published on the Internet.

217 We have forecasts for 48 hours for the Danube. The forecast is

218 the larger the body of water, the more accurate it is. Because if I have errors in the

219 precipitation forecast, they more or less cancel each other out.

220 For example, if I have a thunderstorm and it is forecast by the ZAMG at 221 by the ZAMG and it is 10 km away, then it will affect

222 affects a completely different river, and where it is forecast

223 , there is nothing, but there is something on the neighbouring river. So a thunderstorm is an

224 extreme example, but it illustrates it quite well. The smaller the catchment area

225 catchment area, the more difficult the forecasts are.

226 The forecast models calculate with flow rates and not with water levels.

227 Because a flow rate applies everywhere and a water level only applies at one 228 profile. If the profile is wide, the water level is lower. The flow rate is always the same and independent of the profile. However, if the conversion from water level to flow rate does not work for some reason, then there will of course be problems. 230 231 232 233 234 235 236 237 238 239 240 241 242 243 244 245 246 247 248 249 250 251 252 253 254 255 256 257 258 259 At this water level, we have this flow rate. If the river bed is deepened during a flood, then I naturally have much more flow at this water level. Or, conversely, if something is deposited, then I have less at this water level. This means that these water level keys must also be constantly updated. In this case, one must assume that this is a known uncertainty factor.

235 less flow at this water level. This means that these level keys must also be constantly updated. In this case, one must assume that this is a known uncertainty factor, but that is just the way it is. This means that measurements are always taken at the gauges at as different water levels as possible, a key curve is also created, water level-flow rate, and then a mathematical relationship between water level and flow rate is established.

242 And that is as accurate as possible. We always measure water levels and then use a water level-flow relationship, or so-called key curves, to calculate the flows.

For forecasts on the Internet, we have the water level curve of the measurements, the dark blue line, and then we have this deterministic or best forecast and then, in grey, the 10% / 90% percentile from the precipitation forecasts. There is no hydrological uncertainty here, but it does actually exist.

With regard to the power plants, we have also tried to simulate the operational management on the Danube. We do not have any direct contact there, but there are rules that must be followed. Or even polder systems. When is this polder system flooded to retain the high water peak and relieve the pressure?

I: I thought that might work throughout Austria!

E2: Well, as I said at the beginning, how it has developed and each hydrographic service only covers its own federal state. I don't want to say here how far that makes sense.

I: Where are these measuring stations located?

E2: On the one hand, you can see them in the "Lower Austria Atlas". On the other hand, the ones we use are on the internet. We have an overview map on our homepage. I can then show you in practice afterwards

I can show you afterwards how it actually looks in practice. I: How many of these are currently in use?

I: How many of these are currently in use?

E2: All of them.

I: What does a level measuring station look like?

E2: So we have an old tape recorder inside that basically writes it down with a pen and then the electronics with the remote data transmission. The only absolute thing is really just the level bar that you can read, because we normally measure our water levels with so-called pressure probes. That means it measures the water pressure above the sensor. First of all, they have a certain measuring range where they are calibrated. If you're above or below that, it's no good. If it freezes, it can break anyway. If it gets dry, it can break, or if it gets wet again, it can break. , it can break, or if it gets wet again, it can bad. It also has a certain drift. So a level gauge is only absolute value. You don't know what we're getting, whether it's right right. You sit there and get a value. But the values are, let's say okay, but there's always a margin of error of a few centimetres, and if things go really badly, they can be quite wrong. But that 's quite rare. And you notice it. Typical mistakes are that in winter it freezes and then you get some crazy values, usually insanely high high. Or after a flood, the probe gets muddy, either slow down or stop working altogether. . So those are the typical errors. So that goes back to the the fact that the devices are quite maintenance-intensive. We have relative level values. 1 metre is below the river bed, which means that has nothing to do with the water depth, usually.

I: How maintenance-intensive are these services?

E2: Very. We have a maintenance team that is constantly on the road. It's not like you switch the measuring points on once and then they run forever. They need a power supply. They need some kind of data transmission unit; those are the most important points. If the power supply is down, nothing works, we don't get any data and no one knows why. That means you have to go there in any case. It may be that the power is really just out, that a fuse has blown, it could be anything, or it could be a more serious problem. We receive data, but we don't know if it's correct. All of these electronic devices have measurement errors. Let's assume that they are correct . However, there are so-called measurement point observers on site, who are part-time employees. Essentially, they get a little money for checking once a week, or daily in the case of precipitation measuring stations. They write down the measured values by hand, and these are then compared. If there are significant differences, we have to go there and see what's going on. Some of these part-time employees also check the manual measuring stations. So there is a separate measuring station observer for each measuring station. And there are different types of measuring stations. Some have electronic equipment, some don't, and some have more.

I: Are there different water levels?

E2: As you can see on the Internet, we basically use the hydrological classification. We have average water flow, annual high water levels, and then we have five-year and 30-year levels. This is a hydrological classification so that you know roughly where you stand.

But that has nothing to do with the warning levels, the hydrological classification. Because these warnings depend on the facts, on when the river actually floods. And there are rivers, unfortunately still, where even below HQ 1, I'm thinking of the Danube, where some people are affected. Be it that they can no longer drive to their second homes, or whatever. And that's exactly why 322 these flood warning levels are significantly below these hydrological ones.

323 The hydrological limits are not warning levels per se, but facts. This has nothing to do with whether someone needs to be warned or not. We already have an internal alarm manager, which is also software on a server. All these warning levels are already stored there and it automatically sends out messages, but only to the state warning centre, and we also distribute them internally.

329 I: Are there any countermeasures when an early warning is triggered?

330 (Closing locks, sandbags, alerting emergency services)

332 E2: The state warning centre is a hub for passing on warnings. In very large flood situations, the management team then meets at the state warning centre, but they only coordinate. The state warning centre knows who to notify at which water level.

336 It's similar to the fire brigade control centre, which doesn't do anything itself, it just tells them to go out. The state warning centre is similar. The state warning centre has all the alarm plans, but it basically looks at its list and says, "We've now received a warning. The Kienstock gauge, which is located on the Danube, has exceeded 615 cm. They then look at their list, which they probably already know by heart, but they check

343 at 615 cm, the district administration in Krems and the city of Krems must be notified, and perhaps also the district administration in Tulln. And 345 that's what the state warning centre does. So it has very specific contact persons. Actually, the municipality is responsible for implementing the measures. Again, it's not the district administration, but the district administration then continues in its area. The municipality must then have certain special alarm plans in place, okay, water level 615 cm, what do we do? Do we start setting up flood defences? Or do we drive around in a loudspeaker van and tell people it's time? That's all in the plans. And the local authorities carry it out. But the local authority has usually outsourced this to the fire brigade.

355 But the mayor is the disaster manager on site. The whole thing ends up with him.

357 I: Who has access to your results/who are the users of the

358 results?

359 E2: They are completely accessible to the public.

360 I: Is the data obtained sold?

361 E2: We do not sell data as a matter of principle. It is available to the public

362. We have data on the internet, but you can currently call us and say: "Please, I would

like the data from gauges 1, 2, 3 for the last 20 years, or as far back as it goes." That has nothing to do with the flood thing. Why? We are a public service and we are actually paid for it. We do not sell the data. What we do is, if you say you want the entire country, then that is not a small amount. That means we charge for the effort. How should I know if it takes five hours

370 someone has to sit down and do it, and you have to charge for the five hours according to the data records.

372 I: How is the measurement data processed electronically?

373 E2: On the Internet, they always have a time series, i.e. an Excel table for each month, or a graph showing either one year for surface water and precipitation or 15 years for groundwater. 375 You can also find the data elsewhere, namely in eHyd. That is the website of the Ministry. There you can find the data, perhaps not in the same resolution as you have it, but it is available.

376 You can also find the data elsewhere, namely in eHyd. That is the Ministry's website. There you can download the data directly, although perhaps not in the same resolution as we have. You also have to understand the data. What we publish there is raw data. If there are errors, there are errors. The whole thing then goes through a process, a so-called consolidation, which always takes one to two years because a precipitation balance is then calculated. This means that we have a certain amount of precipitation, of which a certain amount seeps into the groundwater and a certain amount runs off, and this must then match the measurements. Once that is complete, which is why it takes so long, we publish it in the so-called hydrographic yearbook. As soon as it is in the hydrographic yearbook, the time series, which are more like tables from the previous year, are published in eHyd for download. There, for example, you will only find the daily average values for the water level or flow rate, and not the five-minute values that we have. If you say you need the five-minute values for any kind of evaluation, you are welcome to call us and we will provide them to you.

Raw data up to the present time, the data is available to us every five minutes, or the data that has already been checked, then checked again two years later, which you can also have in high resolution.

398 I: How do you obtain current water levels and precipitation values?

400 E2: That's how it looks. The measuring points outside send the measured values every five minutes 401 or every 15 minutes. So it depends on what kind of measuring device 402

it is. A water level gauge sends data every five 403 minutes, a precipitation gauge sends data every 15 minutes.

404 I: Are there any differences between the water level measuring devices?

405 E2: We have two companies that supply these, broadly speaking.

406 technology, they are of course different. We know that these are different devices, but for an outsider, that doesn't matter.

409 I: How do you process the measurement data?

410 E2: Everything goes into a large database. So, first into a central database. This database is then accessed by the web application, i.e. the web application retrieves the data from this database and then, let's say from the hydrological database, the data goes into the forecast database. There it is processed further, forecasts are calculated, which are calculated automatically once an hour. Once a day, the system is checked to see whether the data is reasonable, whether the forecasts are reasonable, or whether there are any serious deviations that are not plausible. If not, the automatic forecast is adopted and all further forecasts for that day are then also automatically adopted and published. In the event of flooding

422, no forecast is published automatically. Each individual forecast is checked, because in the event of flooding, the precipitation forecasts naturally change more quickly. You always have to check whether the precipitation forecasts are plausible and stable. If I have a similar precipitation forecast three times in a row, i.e. three times in a row at hourly intervals, I can assume that it is reasonable. If it is very high once, then very low again and then very high again, you have to question the whole thing a little. For this reason, flood forecasts are only published manually and not automatically. This is so that each individual forecast can be reviewed.

433 I: What does the statistical evaluation and regionalisation of the data look like?

435 E2: The statistics are done in our department. The colleague downstairs does the evaluation of HQ 1, HQ 30, or whatever from a specific point in the river. However, he does not convert this into water status or cross-reference it with the terrain. These are such large-scale designations that you find on the Internet; these are things that are assigned. But essentially, our department does the HQ evaluation for the rivers, for the water levels. The blending is calculated externally for these large projects or by the Water Engineering Department of the in individual cases.

444 I: What impact do the decisions you make have?

E2: We have to decide what forecasts we publish. That's not difficult in standard cases. In extreme cases, when it comes down to it, we have two forecasts, one that we believe is better and one where we believe it is worse. And the one we believe is a little worse, is higher and would mean that, for example, a dam would be flooded, for example. It is of course difficult to make this decision. But these are the decisions we have to make. So we have to decide which forecasts we publish and and forecasts are, of course, always subject to inaccuracies. Of course. Or we publish them and say that we are within a inaccuracy of +/- 10 cm, which is already very accurate. Ultimately, , decisions such as evacuations will still be with the mayor. But the decision on which forecast we publish and where we say that this is the most likely scenario we are actually working with actually expect, is ours. What the mayor then does with that is another matter.

I: Do you tend to make short-term or long-term decisions?

E2: Yes, yes. Yes, they are short-term. Although when it comes to forecasts, will only come into effect later in the future, it could be several hours or several tens of hours. When I think about whether to evacuate Stein or not, then you're there anyway when the flood defences are flooded or not, you need hours of lead time anyway.

I: How do you finance your services?

E2: They are financed by the state, yes. Under certain circumstances in cooperation with other companies. We have a cooperation with EVN on the Kamp and on the Erlauf, because they have their power stations there and are therefore also interested in receiving forecasts. We have also jointly financed the Danube model with Upper Austria.

I: Do you also provide reporting services?

E2: We have to produce status reports, but only in the event of flooding. Of course, you can do the whole thing by telephone, of course, but everything that is written down at is permanent. We also publish the whole thing on teletext page 618. We publish data via the voice server, which can be reached by telephone.

17.3 Transcript Expert 3

Interlocutor:	Stefan Kreuzer
Date	18
Time	11:30 a.m. – 1:00 p.m.
Location	Lower Austrian Provincial Government, Landhausplatz 1, 3109 St. Pölten
Participants:	I: Julia Hasenhütl E3: Expert 3

I: To begin with, could you describe how your department is organised?

E3: We are one of the state's specialist departments. The state has 68 or 70 departments, and these departments are organised into groups. These groups are, so to speak, mirror images of the ministries.

We are called IVW. Internal Administration is the fourth department in this group, IVW4, Department for Fire and Civil Protection. The internal administration group basically reflects what is reflected in the Ministry of the Interior. And so the state has a health group, an economy, sport and tourism group, and they have tried to cluster departments that are reasonably related to each other within the state. Our department for fire brigade and civil defence is responsible for all disaster control. Our department for firefighting and civil protection is responsible for all disaster control. We are based on the Lower Austrian Disaster Control Act, which is a state law. All other federal states have similar laws, with slightly different names, but in principle they are regulated in the same way.

Our department is organised so that we currently have 24 people and three divisions. One division is disaster control planning and organisation. The second division is technology, which includes all warning and alarm systems. I'll tell you more about which warning systems are included. The third department is the state warning centre, which is simply a special organisational unit with nine employees who do nothing else but work shifts to man the control centre. 's headquarters is located in Tulln, along with our department. In peacetime, this department is a normal department, but when a disaster occurs, it is reinforced with telephone operators and dispatchers in addition to the State Warning Centre.

27 We also have staff rooms behind it, and this is where the staff work takes place. This is a special organisational procedure, so to speak. You could also call it a special organisational structure, as it is known in technical jargon, and it is based on a guideline from the Ministry of the Interior called the SKKM guideline. This guideline specifies how we deal with disasters in Austria. In other words, it regulates how a staff works. A staff is nothing more than a group of people sitting together, and each member of the staff has a very specific task. This is regulated in the SKKM guideline.

36 And these SKKM guidelines are applied by all emergency services, 37 state authorities, and anyone who has a disaster control agenda. The most important 38 levels are the municipalities, of course, at the lowest level. At the district level, it is the district administration, at the state level, it is the relevant state department, and at the federal level, disaster control is located in the Ministry of the Interior. However, the Ministry of the Interior only has a coordinating function. There is no federal law for disaster relief; it ends at the state level. That's just how it is in Austria.

44 That's right. It's a federal state; the federal states are organised, and I would say, in my personal opinion, that we also need some kind of legal regulation at federal level. So that the federal government can then say: "Well, I have to take action now. It's not so good for one federal state

48 not so good for one federal state and perhaps a little better for another. And that's

49 Because up until now, people have thought that federal states can cope with

50 the floods in our area, but what do you do when many federal states are affected? That's where the problem lies. So that's how the company is roughly organised, with the cooperation of the local authorities, district authorities, state authorities and the Home Office.

54 I: What is your area of responsibility?

55 E3: My responsibilities are, I am in charge of the area, so I am

56 deputy head of this department and responsible for

57 disaster control planning and organisation. This means that I am responsible

58 that this team functions, that the plans are in place and that they are also 59 practised and that we provide training. That means that the majority of 60 my work is spent in the

classroom, teaching people and training people.

61 We organise exercises ranging from small exercises at the municipal or district level

62 to large regional exercises or even cross-border exercises with neighbouring countries.

63 neighbouring countries. That is my main job.

64 In addition, I recently started teaching at the main university, since

65 Dr. Glade started offering this ÖRISK course.

66 E3: We are purely a networking department. A brief example: the topic of the day is the

67 E3: We are purely a networking department. A brief example: the topic of

68 flooding is only one aspect. We also have other disasters

69 , which is why we are here today. When I am in St. Pölten, because we

70 are based in Tulln, I have to take advantage of the fact that I'm there to drop by and ask

75 what's new. So I don't just go to the department where we are at the moment, 72 but I also go up one floor, where the chemists are based, because when it comes to chemical accidents and DIN standards, 74 we work very closely with the chemists. If we have a nuclear power plant accident, we work very closely with the radiation experts, who are also based in this building. If we have landslides, we need the geologists, and thankfully the experts are all based here in this building. All these experts are available to us on call via the state warning centre. And we are nothing more than a hub where all the specialists in the state of Lower Austria are connected by telephone, so to speak. That means we cooperate. I have only given three examples, but this applies to all areas you can imagine.

84 I: What training did you have to undergo to perform your duties?

86 E3: I went to a normal secondary school. I then studied soil science at university and, thankfully, was able to write my thesis on disaster control. My professional career was at Boku until the end of 2002, and I have been working for the state since 2003.

90 I: What is your professional background?

91 E3: It was all a matter of chance and, of course, my qualifications, because at the time I

92 At the end of the 1990s and beginning of 2000, from the university side in the field of natural hazards, 93 was probably one of the few who had a lot of expertise. I was a volunteer, and still am, with the fire brigade, so I have this network, organisational

knowledge from the emergency services and from the scientific side, such as how the processes work. 97 Early warning system. I would like to install a mudslide early warning system so that

98 I can detect mudslides early on. Or even flooding in a 99 small catchment area where I have relatively little time. How do the emergency services find out about it?

100 the emergency services? How do the authorities find out? How are measures 101

implemented? That was certainly a decisive factor at the time.

102 I: What resources and tools do you have at your disposal to fulfil your current tasks?

104 E3: Electronic tools now include the hydrology alarm manager, which also belongs to the hydrology department. We are, so to speak, only the service centre where you can check in 24 hours a day. 107 That means we also have two colleagues who are in charge of this alarm manager at the state warning centre and who are simply the contact persons for hydrology here, so that they also have a counterpart here. Otherwise, the hydrology department here wouldn't know which of the 24 people here they should talk to. What they then decide is what counts. This means that this alarm manager is an important tool for us, and the other tools for forwarding information are, of course, the normal telephone, radio and a pager network that is also operated by the province of Lower Austria. We also send a lot via text message. 116 But the main means of communication is still the telephone. Email 117 is of course used as a supplement. If you have longer messages, 118 such as flood forecasts, written down or formulated, I'll just give you an example, if we were to receive a precipitation warning, that would be the first step. The ZAMG issues a rainfall warning, which is also the severe weather tool available on the ZAMG website. This means that we often receive a standardised email message from meteorologists at the state warning centre one step ahead of time. This allows us to see whether rainfall is expected that could lead to flooding.

125 rainfall is coming that could lead to flooding. That

126 In the second step, we then contact the hydrologists.

128 Of course, they also receive the weather warnings, but it seems that the sun is still shining at 129, so there is no rain in sight yet. But you can already sense it, especially near the Danube, because you have such a long lead time there, something could be brewing. That means that the state warning centre is already in contact with hydrologists and meteorologists, probably several times a day, and then they make an assessment. You are

in an intensive exchange of expertise. The citizens don't see anything yet. And then, to be on the safe side, you also have these gauges that tell you when you have really reached the water level at which you then actually go out and take action. It may well be that we use hydrology and meteorology to produce a written forecast and texts, which we then send out to a large distribution list from the Landeskommandozentrale (state operations centre). This means that the district headquarters receive it and then distribute it to their municipalities. All emergency services receive it. Infrastructure operators such as hospitals in Lower Austria receive this so that they can plan certain personnel matters in advance. Krems Hospital is also at risk during major floods. This means that it is important that they know about this days in advance. And this distribution list is actually sent out regularly. This means that the people on the distribution list can deal with such forecasts and warnings from our headquarters. They know what to expect and rely on this information. As the event approaches, it is quite possible that something will be sent out twice a day. These are texts and forms that are standardised. We use the forms when a warning is issued. However, before that, we may send out a normal email with the information.
152 inserted in a standardised format. We use the forms when it's already heading towards a warning. However, before that, we may send out a normal email saying: "Please note the weather report, it is forecast that we are likely to expect flooding on the Danube. The frequency is not yet clear because the forecasting systems are not showing that yet, but we know it will rain a lot." It is important for the public sector, especially when it comes to weekends, to make people aware that they can also be reached at the weekend. We are now very close to flooding.
161 Now the levels are rising. Now we're getting to the plans. And 162 these are now also important tools for municipalities to have plans 163 that specify where is located and what to do at which water levels. This 164 is now working quite well in the Wachau municipalities, which have mobile 165 flood protection systems because they require very sophisticated logistics plans for their installation. Who brings which beams where and when? The municipalities are well prepared in this regard. In municipalities where the risk of flooding is not so great, perhaps because only a few houses are below the water level, we also have municipalities that do not have plans because they say, "We have enough time anyway." This is also related to the fact that we have a very strong volunteer culture, especially in the fire brigade, and if the mayor has a well-organised team that does this for him anyway, I naturally have less organisational preparation to do. What other

resources do we have? Because I started with the electronic ones, the radio, the warning and alarm system – the civil defence sirens. That means that if a municipality has a problem very quickly, where you can no longer tell the citizens in person to press the siren so that people know what has happened, there are roughly 2,400 sirens in Lower Austria, which can also be triggered by the state warning centre.

179 know what has happened. There are roughly 2,400 sirens in Lower Austria, 180 which can also be triggered by the state warning centre. But they can also be triggered by the municipality. It may be that the radio connection from Tulln is not working, and then the mayor can also press the siren button. It's such an important tool. Then we move on to assessing the situation. That's a technical term. When I find myself in a flood like this, I always have to assess what it means. Of course, the tools help us here, on the one hand the forecasting systems of the hydrology department, and on the other hand, upstream, the weather warnings from the ZAMG. This means that all the homepages are then 189 glowing when they are in use. They are also equipped by the web servers in such a way that they can withstand high access numbers. 2013 showed that the 191 systems ran largely stable.

192 I: What sources of information do you use on a daily basis?

193 E3: The ZAMG warning page and verbal information from meteorologists and 194 hydrology. Information sources are also available outside. This hierarchy is municipality-district-state. The hierarchy is also established among the emergency services, especially the fire brigade. Staff structures are in place for the emergency services and the same applies at the authority level. The mayor speaks to the district administrator, the district administrator speaks to the county governor.

200 I: Which sources are the most informative and action-oriented?

202 E3: Yes, so these electronic sources, these forecasting systems, 203 precipitation forecasts, are of course very, very important decision-making tools.

205 I: How do you manage this large amount of information?

206 E3: That is ensured by the staff work. There are simply enough people in the staff who process what comes in electronically, verbally, by telephone, radio, in a structured manner, well and thoroughly documented, so that no information is lost. What you also have to agree to there is that disaster response also means chaos. Keeping calm, waiting until the next programme item comes up, what the system says.

210. But disaster response also means chaos. Keeping calm, waiting for the next item on

the agenda, what the system says.

I: How do you get the current water levels?

E3: The internet, and if the internet fails, then from the hydrology department via telephone.

via the telephone

E3: What processes and procedures support your work?

E3: Thanks to our teamwork, there is a very clear workflow who talks to whom, including through the hierarchy of mayor with district > county, which is clearly regulated.

I: Are there different water level stages?

E3: We currently have two on the Danube, specifically for warnings. One is the warning and then the alarm. And that's how it works. The warning is at a certain level. Then everyone everyone calls each other . But no one needs to fill a sandbag or simply , but because a flood is coming. flood is coming. But it could also be that there won't be a flood . The next level is then the alarm, the second value, where active measures must be taken at . And because we have this nomenclature, we have two districts, the district of Tulln and the Vienna area, but the Vienna area will be dissolved next year. In these two administrative areas, it is the case that the water levels for the warning are so low and measures must be taken very, very early on that in these two districts omitted the warning level and immediately issued a high water alarm. Why? Because the entire beach resort settlements in Kritzen- dorf, and there are also beach resort settlements north of the Danube in Tulln, are already being flooded because the settlements are in the flood zone and are flooded even at relatively low flood levels. If you stick to the nomenclature, you would have to set the warning so high that you would probably have a flood warning ten times a year. If you stick to the nomenclature, you would have to set the warning so high that you would probably have a flood warning ten times a year, with water levels that no one would understand. But the people in these beach resort settlements are already used to access roads or footpaths being flooded. But others on the western Danube have no problem with this. Simply because they have such low-lying settlements. However, this is not a permanent settlement area; people know that their stilt houses are in a flood zone. There will never be any flood protection for these settlements.

248 I: Are there any countermeasures in place if an early warning is triggered?

249 E3: Each power plant has very precise operating regulations for how to proceed. We have nothing to do with that. Once the flood alarm is triggered, we start taking countermeasures. Where do I put the sandbags, where do I set them up for the population? That's at the municipal level. Closing sluices. Where you have flood protection dams, you usually also have drains that pass through the dam in case water goes to the other side of the dam, so that it can be left behind. These must then of course also be closed and checked to see if they are functional. So these countermeasures are also included in these plans. But it should be added that not every municipality has these plans.

260 I: What is the purpose of your work?

261 E3: Protection of people, animals and property.

262 I: Who has access to the results?

263 E3: Distribution network with which we are in constant contact.

264 I: What impact do the decisions you make have? 265

266 E3: Impact of decisions. Something really big would have to happen for us to issue an order. According to the Disaster Protection Act, responsibility lies primarily with the district administrator. The mayor is responsible under municipal regulations and legislation, including the Fire Service Act, and must take action in his municipality in any case. The difference between a disaster and everyday life is that the district administrator is, so to speak, the authority who really looks at the district to see where cross-organisational measures need to be taken, where coordination is simply necessary. Where the district administrator may also have to issue orders. At the state level, we mainly coordinate resources that are in short supply, such as helicopters and sandbags, to ensure that they are available from the

278 Foreign countries are concerned. To the point that people in Brussels are saying we need something. But 279 there's also the other side of the coin: we also offer aggregates, for example. 280 Key decisions are made at the state and district level. 281 According to the principle of subsidiarity. The aim is to solve the problem where it arises, which I think is very, very smart, rather than three levels above. The level above only helps when the level below says it can't do any more.

285 I: What measures do you see to improve the effectiveness and efficiency of your services?

287 E3: We are relatively happy with the Danube. Our focus is now on understanding the process of the Danube, i.e. how flooding occurs on the Danube, so that we can communicate this understanding to a larger group of people. But that is the plan for the next two to three years. This will initially result in the compilation of these findings and then probably in training concepts for evening seminars where fire brigade commanders and interested Danube residents can learn more about their waterways. This will be more in communities with large floodplains, where the Danube discharges water into floodplains, in order to understand these processes. The main thing is that people don't forget what they've learned.

298 I: Do you tend to make short-term or long-term decisions?

299 E3: Short-term in the event of a disaster. Long-term in normal operations. Radio systems, for example, are for 25 years.

301 I: How do you finance your services?

302 E3: It's part of the state budget.

17.4 Transcript Expert 4

Interviewee:	Kickinger Peter
Date	29
Time	10:00 a.m.
Location	Upper Austrian Provincial Government, Kärtnerstraße 10-12, 4021 Linz
Participants	I: Julia Hasenhütl E4: Expert 4

1 I: What is your specific area of responsibility?

2 E4: I have been with the company for about a year. I was recruited after the three floods in 2013 to work on projects in the field of hydrology. I am specifically responsible for data transfer and for setting up the database centre. The main focus is on optimising, improving and operating the flood forecasting system. The flood service itself is the responsibility of the head of surface water. My position, on the other hand, is not to manage this area, but to provide support and project management.

11 I: Could you describe how the department is organised?

12 E4: Breaking it down from the state service to the department, it is the Surface Water Management Department. This is actually structured a little differently in every state. We

are now in the Environment Department. The group under this department is the Hydrography Group. Hydrography, or the hydrographic service, which is probably a sub-division in every state, exists in every state. There are nine hydrographic services plus VIADONAU, as a separate hydrographic service, and MA 31 in Vienna, which is responsible for water supply. And in Vienna, there is also the Ministry, which is responsible for hydrography.

separate hydrographic service, and MA 31 in Vienna, which is responsible for water supply. And in Vienna, there is also the ministry, the central hydrographic office. That means it is the headquarters of the hydrographic services. In our case, we are responsible for flood forecasting and flood warnings in surface waters.

25 I: To what extent do you cooperate with others?

26 E4: Specifically for the forecasting systems, we have an annual meeting, an

27 It's called an experience exchange, where we share our experiences with each other and try to learn from them, because it's actually only 29 the that run these forecasting systems. Otherwise, it's a bit difficult to gain experience in other areas. Otherwise, we do it internationally, which happens quite often. We also have meetings, which happen quite often, with the HND Bayern, the Bavarian flood information service.

34 I: What training did you have to undergo to perform your duties?

36 E4: My training was at the University of Natural Resources and Applied Sciences, where I studied cultural engineering and water management. I specialised in hydraulic engineering and hydrology.

39 I: What is your professional background?

40 E4: After graduating, I worked at VIADONAU for nine years and spent the last few years there as head of hydrology, as it is called in that case. Hydrology and hydrography are just different names, and I was head of the flood service for this area. I imagine you've heard of Lower Austria and the Danube flood competition.

46 I: No.

47 E4: VIADONAU is the managing body of the Danube flood control authority. That covers approximately 350 km of dykes from Lower Austria, from Ybbs to the border with Slovakia, and VIADONAU, as the dyke operator, is now also responsible for flood control. This means that there are outposts along the dams and, as with a flood protection association, this work is then carried out. However, the hydrographic services, i.e. the federal states, are responsible for flood forecasting. In this case, VIADONAU has a

special status because it acts like a flood protection association

56 and operates the dams. It takes over flood forecasting from the federal states, makes its own assessments and can then also operate the dam.

59 It's like the Marchland Dam Company. They also operate a large number of dam systems and, so to speak, have their own assessments and operate them independently.

62 I: What resources and tools do you have at your disposal to fulfil your current tasks?

64 E4: We use the water level measuring stations on the Danube and its tributaries as aids. If we now mainly go by the principle of high water forecasting, which we also carry out on the Danube and on the Inn and partly on the Steyr, the entire catchment area is relevant on the Danube. So, the water levels we monitor and the water levels in Salzburg and Bavaria are all relevant because the smaller the catchment area, the fewer areas are affected.

71 So all those in the relevant catchment area.

72 I: Which of these sources are the most informative and decisive for action?

74 E4: The most relevant are the precipitation data. The precipitation forecasts we receive are mainly from ZAMG, 76 where we get precipitation forecasts and analyses. That means the precipitation that has fallen and is forecast. The precipitation then enters the hydrological system and, if we now look at the Danube model, from the Achleithen area, i.e. Passau, the data is also fed into a hydrodynamic model. This means that the precipitation that has fallen is converted into runoff. So at 82, the Danube enters a hydrological model and then a hydrodynamic model. The hydrological model is relevant in the catchment area. This means that the precipitation that has fallen is the input, and at this point I get a flow value. Then a certain runtime is added until the water actually reaches the Danube. From there, it is then calculated further using a hydrodynamic model.

88 Forecast models look the same from a systematic point of view; the tributaries are all built into a hydrological model and the Danube itself actually runs as a hydrodynamic model.

91 I: How do you manage the large amount of information available to you?

93 E4: In principle, the information is almost always the same as in the Nor-

94 malbetreib, i.e. in dry weather. So I would say that if there is flooding, no more information comes in. This means that we have a daily news service that is carried out. A hydrographic employee sits at the computer or in our news centre every day from 6:30 a.m. and

records all the information. They assess the situation, how the situation is developing, and then certain measures are taken accordingly. This means that we have certain responsibilities and these are precisely defined in alarm plans. The alarm plans are drawn up in advance based on experience and they specify exactly what needs to be done. Once the Danube reaches a certain level, a preliminary alarm is triggered. For the most part, only water level values are relevant. This is because it has been determined that flow values can only be reconstructed after a certain amount of time. A flow rate is always indicated at a power station, but it is only indirect. I can measure the water level and see it. Then I know exactly what the situation is. Of course, there is a flow rate in the background that tells us what will happen next. The water level can always change again because the riverbed deepens or things like that. But within a certain range, the water level is decisive for flood protection, and that's what matters.

115 We normally issue a preliminary alarm, so the BHs are already informed in advance. Then we assess the situation and, depending on how things develop, a team is called together, normally at level one, i.e. when the preliminary alarm is exceeded, level one is exceeded.

120 I: How maintenance-intensive are these services?

121 E4: As with any measuring device, checks have to be carried out. There are
122 different systems with advantages and disadvantages that we have. This means that the water level can be measured with a pressure probe. This means that the device measures the water pressure above the device with a pressure sensor, air is pumped in and the counterpressure is measured.

126 The advantage of a pressure probe is that it requires less maintenance. 127 The advantage of the bubble sensor, on the other hand, is that it is protected from lightning. This is because with the bubble sensor, all the electronics are in the water, whereas with the pressure probe, only plastic pipes go in, and the data collector, for example, is not

destroyed. And it is always the case that regular checks are carried out on site. This means that you have a level gauge, then the level reader or the responsible technician drives there, checks it, reads the water level, and then looks inside the level house to see how the digital probe is recording the total. If the two don't match, we recalibrate the probe or adjust it to the measured water level. Part of the Danube, which is also very difficult during floods, is the wave motion, which can fluctuate by 20-30 cm. And then you have to filter out the waves in that sense and see what the average is in the area.

140 the area. Depending on the relevance of a level and whether a level observer is on site, these measuring points are checked in very different ways. This may mean that a reading is taken every day at 7:00 a.m. and the value is sent to the control centre, where it is checked to see if everything is OK. However, it may also happen three times a week. In addition, there is also a special technician who knows how to use the measuring equipment and who comes by at least once a month to check everything. If there are any deviations beforehand, the level observer can notify the technician. Level observers are employed by the state. The level measuring stations belong to VIADONAU. However, as we are part of a network of hydrographic services, this is not an issue. We have still installed some redundant systems, which is what I predicted. We

We have a pressure probe and a bubble sensor, plus different types of remote transmission, via radio, GSM or dedicated line, so you can transmit the data however you want. And then you look to see if you can achieve redundancy. That means that if there is a failure, you can get the data back in via another route.

158 This means that a measuring station usually has several sensors and multiple transmissions. On the Danube, the measuring stations belong to VIADO-NAU, which also collects the data. Only in a flood forecast are the data processed by the hydrographic service . VI- 162 ADONAU is the hydrographic service on the Danube, i.e. for normal water cycle surveys, where the water balance in Austria is compiled. VIADONAU is responsible for this. The state is then responsible purely for flood forecasting or prediction.

VIADONAU has also distanced itself from this because low and medium water levels are also relevant for shipping.

168 I: What measuring stations do you have for forecasting?

169 E4: Before the floods in 2013, 24-hour forecasts were standard. Now we have moved to 48 hours due to the population's desire for more information. However, the further into

the future a forecast is, the less accurate it is. The closer we get to the peak event, the better we can predict it. Only the 48-hour water levels are relevant for the forecast. There are currently 90 water level gauges.

175 I: Are there any countermeasures in place if an early warning is triggered?

176 E4: After the floods in 2013, the municipalities and district authorities were instructed to draw up such alarm plans. The one relevant to us is on the Danube, where there is a general alarm plan.

177 were instructed to draw up such alarm plans. The one on the Danube is relevant for us, as there is a general alarm plan

179 I: How many of these are currently in use?

180 E4: We would need to compile statistics on the number of hours of failure of the 181 gauges. Some gauges are certainly more susceptible than others because the equipment is older or defective, but normally we have 99% data reliability. 183 Duc to the redundancy, the most important gauges are often equipped with two gauges. What is difficult from an external perspective is that a website where data is processed has some kind of backup system. This is due to redundancy, as the most important levels are often duplicated. What is difficult from an external perspective is that a website where data is processed may require additional steps, which means that some data is not publicly available. However, almost 100% of the data is available for the forecasting system.

188 I: Who has access to the results?

189 E4: The forecast results themselves are made available to everyone on the Internet. The forecasting system itself, i.e. the content, can also be interpreted by the user, but only those that are easy to interpret are published, not the others. Those that cannot be interpreted can only be accessed by the emergency services as additional information.

195 If something seems strange to us, we ask the ZAMG whether their forecast is heading in a different direction.

197 I: How do you process the measurement data and how is it prepared for IT purposes?

199 E4: The model doesn't really have a name; we call it the Danube forecast model because it has so many different components.

201 These are the individual measuring points that come in from outside. The 202 data is transmitted to us via modem, GPS or radio. The 203 data then ends up on a server. The data is then retrieved by a data centre 204, processed there and sent to a data processing 205 programme. This is also the data that is available on the Internet as a time 206 series.

So it is then published on the web

207 or fed into the forecasting system. The data is stored in a central location 208 and, once available, the forecasting system 209 has a control system in the background, which is provided by the company Deltares

210 "Views". This is one of the most widely used systems in Europe.

211 The forecasting system has its own recalibration integrated in the background.

212 If the measured value does not match the calculated value in the forecast model, the system corrects itself. This is a separate system that runs in parallel. The system has been checked and improved step by step.

213 model, the system corrects itself. This is a separate system that runs alongside. The system has been gradually checked and improved.

216 I: Do you have a hydrological information service?

217 E4: We provide information about the situation and how it is developing.

218 developing. We also write flood reports and

219 expert reports. In these, we write an assessment of the situation plus the current and

220 forecast water levels. It's one or two pages long. Anyone can

221 get the necessary information from it. Additional information can also be requested from the hydrographic service.

222 from the hydrographic service.

223 I: How are the warning levels divided?

224 E4: According to the alarm plans, which is also the checklist. Where there are dams

225 are present, on the Danube also with mobile walls, this has been worked out

226 worked out in a project. You always calculate backwards. How

227 long do I need to set it up, and depending on that, the warning is issued.

228 This is stored in the alarm plans and we are responsible for

229 to issue the warning in good time. In addition, in the province of Upper Austria, 230 for smaller and medium-sized catchment areas, the limit values are sent to us by the 231 district authorities (BH) , and information is sent via text message to the responsible parties.

232.

233 I: What effectiveness and efficiency measures do you see for refining your

234 services?

235 E4: The systems in Upper and Lower Austria run in parallel. The information

236 are almost identical and the only synergy effect would be on the Danube.

237 All other waterways must be viewed separately. The cooperation

238 simply has to be intensive and even more intensive than it is now.

239 I: Do you tend to make short-term or long-term decisions?

240 E4: 99% of our decisions are long-term because they have to be so secure

241 that no individual can make the wrong decision in the event of a flood.

242 in the event of flooding.

243 I: How do you finance your services?

244 E4: The water levels are set by VIADONAU, but VIADONAU is part of the Federal Ministry, which is responsible for water rights.

245 Ministry, which has been assigned the tasks under water law

246. Basically, we are financed by the federal and state governments. This is because

247 the Ministry of Life on the one hand and the Ministry of Transport on the other,

248 on the Danube. It is often not such a bad thing that several organisations are responsible for this, because the scale of the problem is too great for any one organisation to handle on its own.

249 are responsible, because in terms of scale, it is hardly possible for one organisation to organise everything.

250. When it rains in Bavaria, our forecasting systems immediately take this into account by increasing the forecast.

251

More Books!

yes

I want morebooks!

Buy your books fast and straightforward online - at one of world's fastest growing online book stores! Environmentally sound due to Print-on-Demand technologies.

Buy your books online at
www.morebooks.shop

Kaufen Sie Ihre Bücher schnell und unkompliziert online – auf einer der am schnellsten wachsenden Buchhandelsplattformen weltweit! Dank Print-On-Demand umwelt- und ressourcenschonend produzi ert.

Bücher schneller online kaufen
www.morebooks.shop

info@omniscriptum.com
www.omniscriptum.com

OMNIScriptum

FSC
www.fsc.org
MIX
Papier | Fördert gute Waldnutzung
FSC® C083411

Kolibri 360 GmbH
Ferdinand-Jühlke-Straße 7
99095 Erfurt, Deutschland
produktsicherheit@kolibri360.de

Druck:
CPI Druckdienstleistungen GmbH
im Auftrag der
Kolibri 360 GmbH
Ferdinand-Jühlke-Str. 7
99095 Erfurt